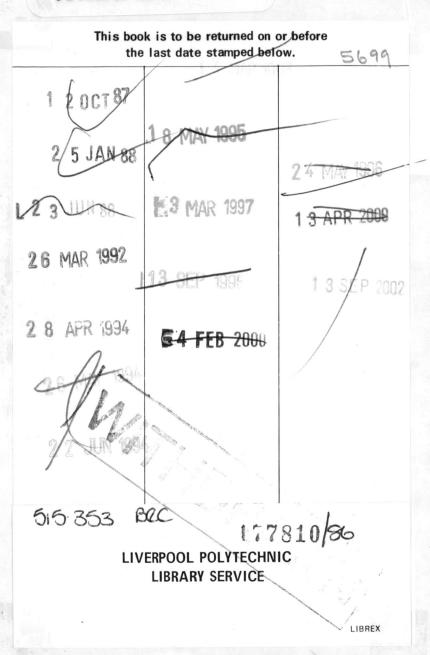

This book is to be returned on or before
the last date stamped below.

5699

1 2 OCT 87

2 5 JAN 88

1 8 MAY 1995

2 4 MAY 1996

2 3 JUN 88

3 MAR 1997

1 3 APR 2000

26 MAR 1992

13 SEP 1996

1 3 SEP 2002

2 8 APR 1994

4 FEB 2000

2 2 JUN 1994

FINITE ELEMENTS
An Introduction

VOLUME I

FINITE ELEMENTS
An Introduction

VOLUME I

ERIC B. BECKER, GRAHAM F. CAREY, and **J. TINSLEY ODEN**

Texas Institute for Computational Mechanics
The University of Texas at Austin

PRENTICE-HALL, INC., *Englewood Cliffs, New Jersey 07632*

Library of Congress Cataloging in Publication Data

Oden, John Tinsley
  Finite elements.

  Bibliography: p.
  Includes index.
  CONTENTS: v. 1. An introduction.
  1. Finite element method. I. Carey, Graham F.,
joint author. II. Becker, Eric B., joint author.
III. Title.
TA347.F5033   515.3′53   80-25153
ISBN 0-13-317057-8 (v. 1)

Editorial production supervision
and interior design by: James M. Chege

Cover design by: Edsal Enterprises

Manufacturing buyer: Joyce Levatino

Printed in the United States of America

10 9 8 7 6 5 4

PRENTICE-HALL INTERNATIONAL, INC., *London*
PRENTICE-HALL OF AUSTRALIA PTY. LIMITED, *Sydney*
PRENTICE-HALL OF CANADA, LTD., *Toronto*
PRENTICE-HALL OF INDIA PRIVATE LIMITED, *New Delhi*
PRENTICE-HALL OF JAPAN, INC., *Tokyo*
PRENTICE-HALL OF SOUTHEAST ASIA PTE. LTD., *Singapore*
WHITEHALL BOOKS LIMITED, *Wellington, New Zealand*

To our children:
Allison, David, Marjorie, and Elizabeth
Varis and Tija
Walker and Lee

CONTENTS

PREFACE

Our purpose in writing this book is to provide the undergraduate student of engineering and science with a concise introduction to finite element methods —one that will give a reader, equipped with little more than calculus, some matrix algebra, and ordinary differential equations, a clear idea of what the finite element method is, how it works, why it makes sense, and how to use it to solve problems of interest to him. We imposed on ourselves three constraints that we felt were of fundamental importance in designing a text of this type.

First, the treatment should not be burdened with technical details that are best appreciated by a more experienced reader. For instance, we feel that discussions of the many variants of finite element methods, detailed aspects of computational schemes for implementing these methods, and numerous applications to problem areas in which the student may have little or no interest are not appropriate in a first course on the subject at this level. Here, we present the method in a form in which the truly salient features can be exposed and appreciated. We choose to relegate those other special topics to later, more advanced volumes.

Second, we did not want to produce either a cookbook or a handbook on finite elements. Although we do give ample coverage of the operational side of finite elements, we also seek to clarify and explain the basic ideas on which these methods are founded. Without these, the student has little foundation on which to build a deeper understanding of either these concepts or their generalizations and, equally important, cannot apply the methods intelligently to difficult problems.

Finally, the book is not aimed at a specific and narrow area of application. The finite element methods are, after all, methods for solving boundary-value problems. Why should a student of, say, heat transfer or fluid mechanics be forced to master structural mechanics in order to learn something about finite elements? We are particularly sensitive to this point because the first course on this subject that we teach is populated by students with such diverse backgrounds and interests as geology, chemical engineering, mathematics, physics, civil engineering, nuclear sciences, aerospace engineering, petroleum engineering, and computer science.

We have each been working on finite element methods for nearly two decades, and this book has evolved as a result of our collective experience in teaching and studying finite elements during this period. Our experience of several years in teaching finite element methods has shown that problem solving and writing and use of simple finite element computer programs is the surest path toward understanding the method. The many exercises and programming assignments should occupy a significant part of the students' time during a semester's study from this book. We strongly believe that this time will be spent to good advantage.

The exercises vary considerably in effort required and in significance. Some of the exercises reinforce, through specific examples, ideas set forth in the text. Others extend the textual material and, in some cases, introduce concepts that, although fundamental in nature, are not necessary to an introductory treatment.

We thank our colleagues and students who have contributed to our understanding and presentation of this subject. We are particularly grateful to David Hibbitt, Linda Hayes, and Gilbert Strang who read the entire manuscript and made many helpful suggestions. We also express our appreciation to B. Palmer for typing the first draft of the manuscript and to N. Webster for assisting with revisions.

E. B. BECKER
G. F. CAREY
J. T. ODEN

Austin, Texas

THE TEXAS FINITE ELEMENT SERIES

1

A MODEL PROBLEM

1.1 ORIENTATION

The finite element method* is a general technique for constructing approximate solutions to boundary-value problems. The method involves dividing the domain of the solution into a finite number of simple subdomains, the finite elements, and using variational concepts to construct an approximation of the solution over the collection of finite elements. Because of the generality and richness of the ideas underlying the method, it has been used with remarkable success in solving a wide range of problems in virtually all areas of engineering and mathematical physics.

Our aim in this chapter is to give a brief introduction to several fundamental ideas which form the basis of the method. For this purpose, we confine our attention to the simplest, most transparent example: a one-dimensional, "two-point" boundary-value problem characterized by a simple linear ordinary differential equation of second order, together with a pair of boundary conditions. We shall refer to this example as our "model problem." Although the model problem is neither difficult nor of much practical interest, both its mathematical structure and our approach in formulating its finite element approximation are essentially the same as in more complex problems of

---

* Throughout this volume, we refer to "the finite element method," as if there were only one. There are, in fact, a variety of methods that employ an element-by-element representation of the approximate solution. Several of these are discussed in Volume II of this series.

greater significance. At many places in this chapter, we pass lightly over points of some practical and theoretical complexity, postponing until later a more thorough treatment.

1.2 THE STATEMENT OF THE MODEL PROBLEM

We begin by considering the problem of finding a function $u = u(x)$, $0 \leq x \leq 1$, which satisfies the following differential equation and boundary conditions:

$$\left.\begin{array}{cc} -u'' + u = x, & 0 < x < 1 \\ u(0) = 0, & u(1) = 0 \end{array}\right\} \qquad (1.2.1)$$

Here the primes denote differentiation with respect to x ($u'' = d^2u/dx^2$). A problem such as this might arise in the study of the deflection of a string on an elastic foundation or of the temperature distribution in a rod.

The *data* of the problem consist of all the information given in advance: the domain of the solution (in this case, the domain is simply the unit interval $0 \leq x \leq 1$), the "nonhomogeneous part" of the differential equation (represented by the given function $f(x) = x$ on the right-hand side), the coefficients of various derivatives of u (in this case these are the constants -1 and $+1$), and the boundary values we demand the solution attain (in this case, zero at $x = 0$ and at $x = 1$).

The data in our model problem are "smooth"; for example, the right-hand side $f(x) = x$ and the coefficients are differentiable infinitely many times. As a consequence of this smoothness, there exists a unique function u which satisfies the differential equation at every point in the domain as well as the boundary conditions. In this particular example, it is a rather simple task to determine the exact solution to (1.2.1), $u(x) = x - (\sinh x/\sinh 1)$. However, in most technical applications, one or both of these happy features of the problem are missing—either there is *no solution* to the classical statement of the problem because some of the data are not smooth, or if a smooth solution exists, it cannot be found in closed form due to the complexity of the domain, coefficients, and boundary conditions.

As an example of the first kind of difficulty, suppose that instead of $f(x) = x$ being given as part of the data (the right-hand side of (1.2.1)), we have the problem

$$-u'' + u = \delta(x - \tfrac{1}{2}), \quad 0 < x < 1; \qquad u(0) = 0 = u(1) \quad (1.2.2)$$

where $\delta(x - \tfrac{1}{2})$ is the *Dirac delta*: the unit "impulse" or "point source" concentrated at $x = \tfrac{1}{2}$. The fact is that $\delta(x - \tfrac{1}{2})$ is not even a function but is rather a symbolic way of describing operations on smooth functions defined by*

$$\delta(x - \tfrac{1}{2})\phi(x) = \phi(\tfrac{1}{2})$$

for any smooth function ϕ satisfying the boundary conditions. We can convince ourselves that if any function u is to satisfy (1.2.2), then it must have a discontinuity in its first derivative u' at $x = \tfrac{1}{2}$; its second derivative u'' does not exist (in the traditional sense) at $x = \tfrac{1}{2}$ (see Exercises 1.2.3 and 1.2.4).

Something appears to be amiss! How can a function u satisfy (1.2.2) everywhere in the interval $0 < x < 1$ when its second derivative cannot exist at $x = \tfrac{1}{2}$ because of the very irregular data given in the problem?

The difficulty is that our requirement that a solution u to (1.2.2) satisfy the differential equation *at every point* x, $0 < x < 1$, is too strong. To overcome this difficulty, we shall reformulate the boundary-value problem in a way that will admit *weaker* conditions on the solution and its derivatives. Such reformulations are called *weak* or *variational* formulations of the problem and are designed to accommodate irregular data and irregular solutions, such as those in problem (1.2.2), as well as very smooth solutions, such as that of our model problem (1.2.1).

Whenever a smooth "classical" solution to a problem exists, it is also the solution of the weak problem. Thus, we lose nothing by reformulating a problem in a weaker way and we gain the significant advantage of being able to consider problems with quite irregular solutions. More important, weak or variational boundary-value problems are precisely the formulations we use to construct finite element approximations of the solutions. We describe such formulations of our model problem in the next section.

Examples of problems for which exact solutions cannot be found explicitly (even though they are known to exist) are found commonly in boundary-value problems in two or three dimensions. It is in the solution of such problems that the true power of the finite element method has made itself felt. The treatment of two-dimensional boundary-value problems begins in Chapter 4.

EXERCISES

1.2.1 Give an example of a physical problem for which the model problem is the mathematical statement.

* The operation $\delta(x - \tfrac{1}{2})\phi$ is sometimes written $\int_0^1 \delta(x - \tfrac{1}{2})\phi(x)\,dx = \phi(\tfrac{1}{2})$ for all infinitely differentiable functions satisfying the boundary conditions $\phi(0) = 0 = \phi(1)$. But even this is incorrect or, at best, only symbolic, because there exists no integrable function that can produce this action on a given smooth function ϕ!

1.2.2 Show that $u(x) = x - \sinh x/\sinh 1$ is the solution of the model problem.

1.2.3 Consider the boundary-value problem

$$-u''(x) = \delta(x - \tfrac{1}{2}), \qquad 0 < x < 1$$
$$u(0) = 0, \qquad u(1) = 0$$

where $\delta(x - \tfrac{1}{2})$ is the Dirac delta corresponding to a point source at $x = \tfrac{1}{2}$. Construct the exact solution u of this problem and sketch u and u' as functions of x. What does the graph of u'' look like? Does the classical statement of this problem given above make sense at $x = \tfrac{1}{2}$? Why?

1.2.4 Construct the solution u of the boundary-value problem (1.2.2) and sketch u and u' as functions of x. Comment on u'' and the meaning of the classical statement of this boundary-value problem.

1.3 VARIATIONAL STATEMENT OF THE PROBLEM

One weak statement of the model problem (1.2.1) is given as follows: find the function u such that the differential equation, together with the boundary conditions, are satisfied in the sense of weighted averages. By the satisfaction of all "weighted averages" of the differential equation, we mean that we require that

$$\int_0^1 (-u'' + u)v \, dx = \int_0^1 xv \, dx \qquad (1.3.1)$$

for all members v of a suitable class of functions. In (1.3.1) the *weight function*, or *test function*, v, is any function of x that is sufficiently well behaved that the integrals make sense.*

In order to describe this weak statement of the problem more concisely, we introduce the idea of the set of all functions that are smooth enough to be considered as test functions. We will denote the set of such functions, which have zero values at $x = 0$ and $x = 1$, by the symbol H. To indicate that a function v is a member of the set H, we use the notation "$v \in H$," which is read "v belongs to H." The variational statement (1.3.1) of our prob-

---

* It is easy to find functions that are not smooth enough to serve as test functions. For example, if $u(x) = x - \sinh x$ and $v(x) = x^{-3}$, then neither $\int_0^1 (-u'' + u)v \, dx$ nor $\int_0^1 xv \, dx$ have finite values and (1.3.1) does not make sense. There is, however, a multitude of functions which are perfectly acceptable as test functions. The exact specification of such functions is central to the theory of the finite element method and will be discussed in detail later.

lem now assumes the more compact form: find u such that

$$\left.\begin{array}{c} \int_0^1 (-u'' + u - x)v \, dx = 0 \qquad \text{for all } v \in H \\[2mm] u(0) = 0 \\[1mm] u(1) = 0 \end{array}\right\} \qquad (1.3.2)$$

Upon reflection, it is clear that, if (1.3.2) is true, there can be no portion of finite length, however small, of the interval $0 < x < 1$ within which the differential equation (1.2.1) fails to be satisfied in an average sense. To see this, we need only hypothesize the existence of such a region and show that, as a consequence, (1.3.2) would not be satisfied. Consider the *residual*, or error, in the differential equation, defined by the function $r(x) = -u'' + u - x$. Suppose that $r(x)$ is different from zero in some small region, such as that shown in Fig. 1.1a. Corresponding to this particular $r(x)$, we can choose $v(x)$ as shown in Fig. 1.1b.* Noting that the integrand in (1.3.2) is positive in the interval $a < x < b$ and zero elsewhere, we see that the integral in (1.3.2) cannot vanish (i.e., (1.3.2) is not satisfied), so that u cannot be a solution of problem (1.2.1). Through various choices of v we can "test" the differential equation in every portion of the region of interest, so (1.3.2) does indeed require that (1.2.1) be true, on the average, over every subregion.

This weak statement of our problem, although seemingly less direct than the classical statement (1.2.1), has a certain appeal for those motivated by physical arguments. In modeling physical phenomena, it is often desirable to measure (or at least to consider the measurement of) the data and/or the solution of a boundary-value problem. Since any real measurement device (strain gauge, thermocouple, etc.) will have finite size, these quantities can, at best, be determined only in some average sense over small regions and not at any particular single point. The weak statement of the problem can be interpreted as assuring us that the solution will appear to be correct when tested at any location in the region with an arbitrarily small transducer.

1.3.1 A Symmetric Variational Formulation

At this stage, there are two points that should be thoroughly appreciated:

1. The weaker formulation (1.3.2) is as valid and meaningful as the original statement (1.2.1); indeed, the solution of (1.2.1) also satisfies (1.3.2) and, in fact, is *the* (only) solution of (1.3.2).

---

* Although we do not give the equation of $v(x)$, it is clear from the sketch that v is smooth enough to serve as a test function.

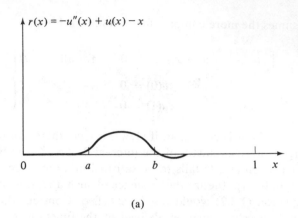

(a)

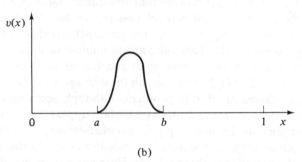

(b)

FIGURE 1.1 *Example of a residual error function* r(x)
= −u″(x) + u(x) − x and a smooth test function v(x). *If* u *is the
solution to* (1.2.1), r *cannot, on the average, be other than zero
on any subinterval,* a < x < b.

2.   The specification of the set *H* of test functions is an essential ingredient
of an acceptable weak formulation.

Let us elaborate on point 2. Although it may not be immediately obvious,
the test functions in variational problems such as (1.3.2) may not belong to
the same class *H* of functions as the class $\tilde{H}$ to which the solution belongs (see
Exercise 1.3.1). The set $\tilde{H}$ to which the solution *u* belongs is called the *class of
trial functions* for such problems. Our smoothness requirements demand that
we consider the pair of sets of functions, $\tilde{H}$ and *H*. For instance, *u* may be
chosen from a class of functions $\tilde{H}$ which have the property that their *second*
derivatives, when multiplied by a test function *v*, produce a function $u''v$ which
is integrable over the interval $0 < x < 1$. On the other hand, no derivatives
of test functions appear in (1.3.2). Thus, even though (1.3.2) is a perfectly

valid weak statement of the model problem (1.2.1), the fact that H and $\tilde{H}$ are not the same leads to a lack of symmetry in the formulation which we ordinarily prefer to avoid. The particular weak form (1.3.2), in other words, is not the most suitable for computational or theoretical purposes. We list some advantages of symmetric formulations in the next section.

We obtain an alternative, symmetric, weak formulation of (1.2.1) by observing that, if u and v are sufficiently smooth functions, then the standard integration-by-parts formula holds:

$$\int_0^1 -u''v \, dx = \int_0^1 u'v' \, dx - u'v \Big|_0^1$$

If we continue to demand that the test functions vanish at the endpoints, then $\int_0^1 -u''v \, dx = \int_0^1 u'v' \, dx$ for all admissible test functions v, and therefore (1.3.2) can be replaced by the following alternative variational problem: find $u \in H_0^1$ such that

$$\int_0^1 (u'v' + uv - xv) \, dx = 0 \qquad \text{for all } v \in H_0^1 \qquad (1.3.3)$$

where H_0^1 is a new class of functions,* the properties of which we shall describe below. Now there is a certain symmetry in the formulation: the same order of derivatives of both trial and test functions appear and we have taken $H = \tilde{H} = H_0^1$. Moreover, we can once again verify that any solution of our model problem (1.2.1) satisfies (1.3.3), so we have still not lost anything in this reformulation. However, since (1.3.2) contains second derivatives of the solution u whereas (1.3.3) has only first derivatives, we see that in passing from (1.2.1) to (1.3.2) to (1.3.3) we have progressively weakened the smoothness requirements on our solution and, thereby, progressively *enlarged* the class of *data* for which these statements of the problem make sense. We shall refer to the particular weak formulation defined in (1.3.3) as a variational boundary-value problem.

Let us now return to the characterization of the class H_0^1 of test and trial functions. The set H_0^1 is also termed the *class of admissible functions for problem* (1.3.3), since it contains only those functions which satisfy the boundary conditions and are sufficiently regular that the integral in (1.3.3) has meaning. The most irregular term inside the integrand in (1.3.3) is $u'v'$: if u and v are irregular, then certainly their derivatives are even less regular. Since v is allowed to be any function in the set of admissible functions, we

---

* In later chapters, we will use the notation $H_0^1(0, 1)$ to describe this class of functions: the superscript "1" signifies that members v of this class have derivatives of order 1 which are square-integrable on the interval $0 < x < 1$. The subscript "0" indicates that $v = 0$ at $x = 0$ and at $x = 1$.

must consider the possibility that $v = u$. Thus, it will be necessary that $(v')^2$ be smooth enough for its integral to be calculated. Functions satisfying this condition are said to have square-integrable first derivatives.* Hence, we shall define the set of admissible functions H_0^1 as the set of all functions that vanish at the endpoints and whose first derivatives are square-integrable. Thus, a function w is a member of H_0^1 if

$$\int_0^1 (w')^2 \, dx < \infty \qquad \text{and} \qquad w(0) = 0 = w(1) \qquad (1.3.4)$$

There is one final point. Even though we have derived the variational formulation (1.3.3) from (1.2.1), it is important that we regard the converse to be true: we shall consider that (1.3.3) is the given model boundary-value problem that we wish to solve instead of (1.2.1). Having solved (1.3.3), we may then ask if the solution is smooth enough to also be a "classical" solution, that is, a function satisfying (1.2.1) at every x in $0 < x < 1$. Clearly, this point of view will make it possible for us to consider broad classes of variational boundary-value problems which have no classical solutions as well as equivalent variational formulations of nice problems with classical solutions.

EXERCISES

1.3.1   (a)   Consider the second-order boundary-value problem,

$$-u''(x) = x, \qquad 0 < x < 1$$
$$u(0) = 0, \qquad u(1) = 0$$

and the following weak boundary-value problems:

(b)   Find $u \in \tilde{H}$ such that $u(0) = u(1) = 0$ and

$$\int_0^1 (-u'' - x)v \, dx = 0 \qquad \text{for all } v \in H$$

(c)   Find $u \in \tilde{H}$ such that

$$\int_0^1 (u'v' - xv) \, dx = 0 \qquad \text{for all } v \in H$$

---

* The condition of square integrability is a more restrictive condition than that of integrability. For example, the function $v(x) = x^{-1/2}$ is integrable over $0 < x < 1$ but v^2 is not integrable over this interval.

(d)   Find $u \in \tilde{H}$ such that

$$\int_0^1 (-uv'' - xv)\, dx = 0$$

for all $v \in H$ such that $v(0) = v(1) = 0$.

Discuss properties of the classes of test and trial functions for each of boundary-value problems (b), (c), and (d) so that these problems correspond to the given problem (a). In addition, suppose that f, g, and h are the three functions shown below:

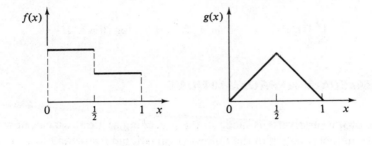

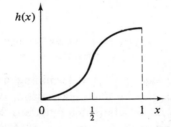

The function h is given by

$$h(x) = \begin{cases} \tfrac{1}{2}x^2, & 0 \le x \le \tfrac{1}{2} \\ \tfrac{1}{4} - \tfrac{1}{2}(x-1)^2, & \tfrac{1}{2} \le x \le 1 \end{cases}$$

For each of problems (b), (c), and (d), determine whether or not f, g, and h belong to either the appropriate class of trial functions or test functions.

1.3.2 Returning to Exercise 1.2.3, show that a proper variational statement of this problem is: find $u \in H_0^1$ such that

$$\int_0^1 u'v'\, dx = v(\tfrac{1}{2}) \qquad \text{for every } v \in H_0^1$$

Verify that the solution u does indeed belong to H_0^1.

1.3.3 Construct a correct variational statement of the problem defined in equation (1.2.2).

1.3.4 Suppose that we are given the variational boundary-value problem (1.3.3) and that we have established that its solution u is smooth. Show that u is then the solution of (1.2.1). How does this exercise differ from that outlined in the text in deriving (1.3.3)?

1.3.5 Show that one variational formulation of the boundary-value problem

$$-xu'' - u' + u = \sin x, \qquad 0 < x < 1$$
$$u(0) = u(1) = 0$$

is to find $u \in H_0^1$ such that

$$\int_0^1 (xu'v' + uv - v \sin x)\, dx = 0 \qquad \text{for all } v \in H_0^1$$

1.4 GALERKIN APPROXIMATIONS

In view of the observations made in the preceding section, we can now consider our model problem in the following variational form: find $u \in H_0^1$ such that

$$\int_0^1 (u'v' + uv)\, dx = \int_0^1 xv\, dx \qquad \text{for all } v \in H_0^1 \qquad (1.4.1)$$

We now take up the question of determining approximate solutions to (1.4.1) [and, therefore, (1.2.1)] and, once again, our approach centers on properties of the class H_0^1 of admissible functions defined in (1.3.4).

There are two fundamental properties of H_0^1, besides those listed in (1.3.4), which play a crucial role in the type of approximation we have in mind. First, H_0^1 is a *linear space* of functions, and second, it is *infinite-dimensional*.

By a "linear space" we simply mean that linear combinations of functions in H_0^1 are also members of H_0^1. In other words, if v_1 and v_2 are arbitrary test functions and α and β are arbitrary constants, then $\alpha v_1 + \beta v_2$ is also a test function.

By "infinite-dimensional" we mean that it is necessary to specify an infinity of parameters in order to define uniquely an arbitrary test function v in the space. The reader with some knowledge of Fourier series will have no difficulty in understanding this concept. Indeed, if we introduce the set of functions

$$\psi_n(x) = \sqrt{2} \sin n\pi x, \qquad n = 1, 2, 3, \ldots \qquad (1.4.2)$$

and v is a smooth test function in H_0^1, then it is easily verified that v can be represented in the form

$$v(x) = \sum_{n=1}^{\infty} a_n \psi_n(x) \qquad (1.4.3)$$

where the scalar coefficients a_n are given by

$$a_n = \int_0^1 v(x)\psi_n(x)\,dx \qquad (1.4.4)$$

Thus, in view of (1.4.3), an infinity of coefficients a_n must be specified in order to define any function $v \in H_0^1$; the space H_0^1 of admissible functions is, therefore, infinite-dimensional.

Let us suppose that we are given an infinite set of functions $\{\phi_1(x), \phi_2(x), \phi_3(x), \ldots\}$ in H_0^1 which have the property that each test function v in H_0^1 can be represented as a linear combination of the $\phi_i(x)$ by a series of the type in (1.4.3). In most cases, we could just as well use the smooth trigonometric functions ψ_n defined in (1.4.2) for this purpose, but we wish to emphasize that the $\phi_i(x)$ need not be trigonometric but may be less smooth functions. Our basic requirement is that each $v \in H_0^1$ be representable as a linear combination of such functions of the type

$$v(x) = \sum_{i=1}^{\infty} \beta_i \phi_i(x) \qquad (1.4.5)$$

where the β_i are constants and the series converges in a sense* appropriate for the space H_0^1 of (1.3.4). A set of functions $\{\phi_i\}$ with these properties is said to provide a basis for H_0^1 and the functions ϕ_i are called *basis functions*.

It is clear that if we take only a finite number N of terms in the series (1.4.5), then we will obtain only an approximation v_N of v:

$$v_N(x) = \sum_{i=1}^{N} \beta_i \phi_i(x) \qquad (1.4.6)$$

The N basis functions $\{\phi_1, \phi_2, \ldots, \phi_N\}$ define an N-dimensional subspace $H_0^{(N)}$ of H_0^1. The subspace $H_0^{(N)}$ is of only finite dimension N because† each function v_N in $H_0^{(N)}$ is determined by a linear combination of only the N functions $\phi_1, \ldots, \phi_N$ by (1.4.6). $H_0^{(N)}$ is a subspace of H_0^1 because each

---

* If v_N is given by (1.4.6), then v_N converges to a function v "in an H_0^1-sense" if

$$\lim_{N \to \infty} \int_0^1 [(v - v_N)^2 + (v' - v_N')^2]\,dx = 0$$

† Here we have assumed that the N functions $\phi_1, \phi_2, \ldots, \phi_N$ are linearly independent; that is, it is impossible to find N coefficients $\beta_1, \beta_2, \ldots, \beta_N$ not all of which are zero, such that $\sum_{i=1}^{N} \beta_i \phi_i(x) = 0$ for all x.

ϕ_i, $i = 1, 2, \ldots, N$, is, by definition, a member of H_0^1. For example, $\{\phi_1, \phi_2, \phi_3\}$ is a basis for a three-dimensional subspace $H_0^{(3)}$ of H_0^1; $\{\phi_1, \phi_2, \phi_3, \phi_4\}$ defines a four-dimensional subspace of H_0^1; and so on.

With these preliminaries behind us, we are now ready to consider Galerkin's method for constructing approximate solutions to the variational boundary-value problem (1.4.1). *Galerkin's method consists of seeking an approximate solution to* (1.4.1) *in a finite-dimensional subspace $H_0^{(N)}$ of the space H_0^1 of admissible functions rather than in the whole space H_0^1.* Thus, instead of tackling the infinite-dimensional problem (1.4.1), we seek an approximate solution u_N in $H_0^{(N)}$ of the form

$$u_N(x) = \sum_{i=1}^{N} \alpha_i \phi_i(x) \tag{1.4.7}$$

which satisfies (1.4.1) with H_0^1 replaced by $H_0^{(N)}$. In other words, the variational statement of the approximate problem is this: find $u_N \in H_0^{(N)}$ such that

$$\int_0^1 (u_N' v_N' + u_N v_N) \, dx = \int_0^1 x v_N \, dx \qquad \text{for all } v_N \in H_0^{(N)} \tag{1.4.8}$$

Since the ϕ_i are known, u_N will be completely determined once the N coefficients α_i in (1.4.7) are determined. The α_i in (1.4.7) are referred to as the *degrees of freedom* of the approximation.

Let us now investigate how we might go about calculating the unknown coefficients α_i. We first observe that all of the test functions v_N are linear combinations of the basis functions ϕ_i of the form (1.4.6), the β_i being arbitrary constants. Note again that v_N in (1.4.6) can take on the values of any function in $H_0^{(N)}$ through a proper choice of the constants β_i.

To determine the specific values, α_i, of these coefficients that will characterize the approximate solution u_N, we introduce (1.4.6) and (1.4.7) into (1.4.8) to obtain the condition

$$\int_0^1 \left\{ \frac{d}{dx} \left[\sum_{i=1}^N \beta_i \phi_i(x) \right] \frac{d}{dx} \left[\sum_{j=1}^N \alpha_j \phi_j(x) \right] + \left[\sum_{i=1}^N \beta_i \phi_i(x) \right] \left[\sum_{j=1}^N \alpha_j \phi_j(x) \right] \right.$$
$$\left. - x \sum_{i=1}^N \beta_i \phi_i(x) \right\} dx = 0 \qquad \text{for all } \beta_i, i = 1, 2, \ldots, N \tag{1.4.9}$$

Expanding (1.4.9) and factoring the coefficients β_i gives

$$\sum_{i=1}^N \beta_i \left(\sum_{j=1}^N \left\{ \int_0^1 [\phi_i'(x)\phi_j'(x) + \phi_i(x)\phi_j(x)] \, dx \right\} \alpha_j - \int_0^1 x\phi_i(x) \, dx \right) = 0$$
$$\text{for all } \beta_i, i = 1, 2, \ldots, N \tag{1.4.10}$$

where $\phi_i'(x) = d\phi_i(x)/dx$.

The structure of (1.4.10) is most easily seen by rewriting it in the more compact form

$$\sum_{i=1}^{N} \beta_i \left(\sum_{j=1}^{N} K_{ij}\alpha_j - F_i \right) = 0 \tag{1.4.11}$$

for all choices of β_i, where

$$K_{ij} = \int_0^1 [\phi_i'(x)\phi_j'(x) + \phi_i(x)\phi_j(x)]\, dx \tag{1.4.12}$$

and

$$F_i = \int_0^1 x\phi_i\, dx \tag{1.4.13}$$

and in which

$$i, j = 1, 2, \ldots, N$$

The $N \times N$ rectangular array of numbers $\mathbf{K} = [K_{ij}]$ is usually referred to as the *stiffness matrix* for problem (1.4.8) for the basis functions ϕ_i; the $N \times 1$ column vector $\mathbf{F} = \{F_i\}$ is referred to as the *load vector* for this choice of basis functions.* Since the ϕ_i are known, the numbers K_{ij} and F_i can be calculated directly by formulas (1.4.12) and (1.4.13).

Because the β_i are arbitrary, (1.4.11) represents N equations to be satisfied by the α_j rather than the single equation it may appear to be. To see this, consider the following natural choices for the set of parameters: $\beta_1 = 1$, $\beta_i = 0$ for $i \neq 1$. Then (1.4.11) yields

$$\sum_{j=1}^{N} K_{1j}\alpha_j = F_1$$

Next, set $\beta_2 = 1$, $\beta_i = 0$, $i \neq 2$, so that

$$\sum_{j=1}^{N} K_{2j}\alpha_j = F_2$$

Continuing in this way, we arrive at the system of N linear equations in the N unknown coefficients α_j:

$$\sum_{j=1}^{N} K_{ij}\alpha_j = F_i, \qquad i = 1, 2, \ldots, N \tag{1.4.14}$$

Since the functions ϕ_i have been chosen to be independent, equations (1.4.14) will be independent, and therefore the stiffness matrix $\mathbf{K}$ will be

---

* The terms "stiffness" and "load" arise from the similarity of equation (1.4.14) to the equation $K\alpha = F$ for the deflection α of a linear spring with stiffness K under an applied load F.

invertible. It follows that the coefficients α_j are uniquely determined by (1.4.14) and are of the form

$$\alpha_j = \sum_{i=1}^{N} (K^{-1})_{ji} F_i \qquad (1.4.15)$$

where $(K^{-1})_{ji}$ are the elements of the inverse of $\mathbf{K}$. The approximate solution u_N is now determined by introducing (1.4.15) into (1.4.7).

Some reasons that the symmetrical variational formulation (1.3.3) of our model problem is preferable over such weak statements as (1.3.2) should now be apparent:

1. Our approximation of the symmetric formulation has led to a symmetric stiffness matrix in (1.4.12); an unsymmetric formulation never will. This symmetry provides us with an opportunity for reducing the computational effort in obtaining an approximate solution.

2. If a symmetric formulation of our model problem is used, it can be shown that Galerkin's method provides, in a certain natural sense, the best possible approximation of the solution u in $H_0^{(N)}$ (see Exercise 1.4.2).

3. For the symmetric formulation, the spaces of trial functions and test functions coincide; hence, only one set of basis functions ϕ_i need be constructed for such approximations.

It is important to note that the quality of the approximation is completely determined by our choice of the functions ϕ_i: once these have been chosen, the determination of the coefficients α_j reduces to a computational matter.

EXERCISES

1.4.1 Show that the error $e(x) = u(x) - u_N(x)$ in the Galerkin approximation of (1.4.1) satisfies the "orthogonality condition,"

$$\int_0^1 (e'v_N' + ev_N)\, dx = 0 \qquad \text{for all } v_N \in H_0^{(N)}$$

1.4.2 Using the results of Exercise 1.4.1, show that if u is the exact solution of (1.4.1), u_N is its Galerkin approximation, and v_N is an arbitrary test function in $H_0^{(N)}$, then

$$\int_0^1 [(u' - v_N')^2 + (u - v_N)^2]\, dx = C + \int_0^1 [(u_N' - v_N')^2 + (u_N - v_N)^2]\, dx$$

where C is a positive constant depending only on the known functions u and u_N. The left-hand side of this equality is a measure of the error between the exact solution u and an arbitrary element v_N of the finite-dimensional subspace $H_0^{(N)}$. Show that the choice $v_N = u_N$ makes this error as small as possible for any given subspace $H_0^{(N)}$. This result shows that, when this measure of error is used, Galerkin's method provides the *best possible* approximation of u in $H_0^{(N)}$.

1.4.3 Returning to the Galerkin approximation (1.4.8), let $N = 3$ and choose $\phi_i = \sin i\pi x$, $i = 1, 2, 3$. Calculate K_{ij} and F_i, solve for the coefficients α_j, and construct the approximate solution u_N. Plot the exact and approximate solutions and comment on the accuracy of your approximation.

1.4.4 Construct basis functions ϕ_i, $i = 1, 2, 3$, that are polynomials of degree $(i + 1)$ and satisfy the boundary conditions of problem (1.4.8). Then rework Exercise 1.4.3 for this choice of basis functions.

1.4.5 Use the polynomial basis functions of Exercise 1.4.4 together with the unsymmetric form of the variational boundary-value problem (1.3.2) to construct the approximate solution u_N.

1.5 FINITE ELEMENT BASIS FUNCTIONS

While Galerkin's method provides an elegant strategy for constructing approximations of solutions of boundary-value problems, it has one very serious shortcoming: in the method as we have described it, there is no systematic way of constructing reasonable basis functions ϕ_i for the approximate test functions v_N. Aside from being independent members of H_0^1, they are arbitrary. The analyst is left with a bewildering number of possibilities at his disposal and with the discomforting knowledge that the quality of his approximate solution will depend very strongly on the properties of the basis functions he chooses. The situation is much worse in two- and three-dimensional boundary-value problems in which the functions ϕ_i must be designed to fit boundary conditions on domains with complex geometries. Moreover, a poor choice of the ϕ_i may produce an ill-conditioned stiffness matrix so that the linear system (1.4.14) may be difficult to solve within acceptable limits of accuracy. For these reasons, particularly the difficulty of treating irregular geometry in higher dimensions, the classical Galerkin method is of rather limited use. These substantial difficulties can be resolved by using the finite element method.

   The finite element method provides a general and systematic technique for constructing basis functions for Galerkin approximations of boundary-value problems. The main idea is that the basis functions ϕ_i can be defined piecewise over subregions of the domain called *finite elements* and that over any subdomain, the ϕ_i can be chosen to be very simple functions such as polynomials of low degree.

To construct such a set of piecewise basis functions, we first partition the domain (i.e., the interval $0 \leq x \leq 1$) of our problem into a finite number of elements. Figure 1.2 shows, for example, the domain of our model problem

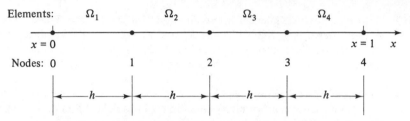

FIGURE 1.2 *A finite element partition of the region $0 \leq x \leq 1$ comprised of four elements with nodes at element endpoints.*

partitioned into four elements denoted Ω_i, $i = 1, 2, 3, 4$. Following a standard notation, the length of each finite element Ω_i will be denoted h_i. Since the elements in the example indicated in the figure are of equal length, we shall denote the element length in this case by h.

Within each element, certain points are identified, called *nodes* or *nodal points*, which play an important role in finite element constructions. In the example indicated in Fig. 1.2, the five nodes are taken to be the endpoints of each element; these are numbered 0 through 4 in the figure. The collection of elements and nodal points making up the domain of the approximate problem is sometimes referred to as a *finite element mesh*.

We introduce a slight change in notation at this point. In the preceding section, we denoted the approximate test functions and solution by v_N and u_N, respectively, where N was a parameter indicating the number of basis functions used in defining $H_0^{(N)}$. In finite element methods, it is customary to use the mesh length h as a parameter instead of N, the idea being that as h becomes smaller, more elements must be introduced and, therefore, more basis functions are furnished for $H_0^{(N)}$. Thus, we shall denote v_N, u_N, and $H_0^{(N)}$ by v_h, u_h, and H_0^h in subsequent discussions.

Having constructed a finite element mesh for our model problem (such as that in Fig. 1.2), we proceed to construct a corresponding set of basis functions using the following fundamental criteria:

1. The basis functions are generated by simple functions defined piecewise—element by element—over the finite element mesh.

2. The basis functions are smooth enough to be members of the class H_0^1 of test functions.

3. The basis functions are chosen in such a way that the parameters α_i defining the approximate solution u_h ($= u_N$; recall (1.4.7)) are precisely the values of $u_h(x)$ at the nodal points.*

One very simple, yet perfectly adequate, set of basis functions satisfying these three criteria is shown in Fig. 1.3a. If the coordinates of the nodes are denoted x_i ($i = 0, 1, 2, 3, 4$), then the basis functions shown for $i = 1, 2$, and 3 are given by

$$\phi_i(x) = \begin{cases} \dfrac{x - x_{i-1}}{h_i} & \text{for } x_{i-1} \leq x \leq x_i \\[2mm] \dfrac{x_{i+1} - x}{h_{i+1}} & \text{for } x_i \leq x \leq x_{i+1} \\[2mm] 0 & \text{for } x \leq x_{i-1} \quad \text{and} \quad x \geq x_{i+1} \end{cases} \tag{1.5.1}$$

where $h_i = x_i - x_{i-1}$ is the length of element Ω_i. Their first derivatives are

$$\phi_i'(x) = \begin{cases} \dfrac{1}{h_i} & \text{for } x_{i-1} < x < x_i \\[2mm] \dfrac{-1}{h_{i+1}} & \text{for } x_i < x < x_{i+1} \\[2mm] 0 & \text{for } x < x_{i-1} \quad \text{and} \quad x > x_{i+1} \end{cases} \tag{1.5.2}$$

To demonstrate that these basis functions satisfy the foregoing criteria, first observe that each of the functions ϕ_i, $i = 1, 2, 3$, is the result of patching together piecewise-linear functions defined over each finite element. For example, the "hat"-shaped function ϕ_1 associated with node 1 is produced by combining a linear function defined on element Ω_1 and a linear function defined on element Ω_2, as indicated in Fig. 1.3b. This illustrates what we mean when say that the basis functions are "generated by simple functions defined piecewise—element by element—over the finite element mesh." We shall see later that the strength of the finite element method rests on this particular way of constructing basis functions. We shall also find that, as a result of such constructions, the approximation (1.4.8) of our problem can be formulated one element at a time, the final formulation being obtained by summing up the correct contributions furnished by each element. These ideas will be clarified in the next section and discussed in greater detail in Chapter 2.

Turning now to criterion 2, we are reminded that in order that $\phi_i \in H_0^1$, $i = 1, 2, 3$, each must have square-integrable first derivatives and must vanish at $x = 0$ and at $x = 1$. The functions shown in Fig. 1.3a obviously

---

* Later we will consider the possibility that some of the α_i may correspond to values of various derivatives of u_h at certain nodal points.

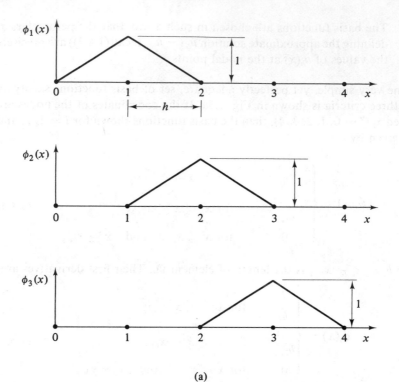

(a)

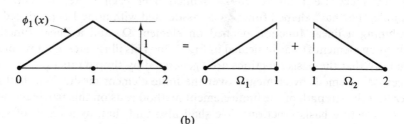

(b)

FIGURE 1.3 *Example of finite element basis functions.*

satisfy the boundary conditions. Are their derivatives square-integrable? The answer is yes, because the derivative of each of the ϕ_i is a step function, of the type indicated in Fig. 1.4a, and therefore, $[\phi_i']^2$ is certainly integrable. Indeed, the integral of $[\phi_i']^2$ is merely the area under the curve indicated in the figure:

$$\int_0^1 [\phi_i'(x)]^2 \, dx = \frac{1}{h^2} 2h = 2h^{-1} < \infty$$

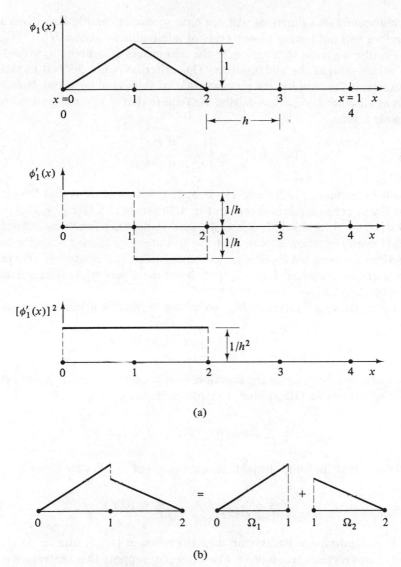

(a)

(b)

FIGURE 1.4 *Finite element basis functions whose derivatives (a) are and (b) are not square-integrable.*

What is critical here is that upon patching together the piecewise-linear functions defined elementwise to form each of our basis functions (as illustrated in Fig. 1.3b), the functions in adjacent elements match perfectly at common nodes. Then the $\phi_i(x)$ will be continuous throughout the domain of the problem. If this is not the case, a function will be produced which suffers a "jump" (a discontinuity) at a node, such as that indicated in Fig. 1.4b.

Such discontinuous functions will not have square-integrable derivatives and therefore will not belong to our class of admissible functions H_0^1.

Finally, we come to criterion 3: the parameters α_i defining u_h should be the values of u_h at the nodal points. This criterion is not difficult to satisfy if each basis function has the property that its value is unity at one node and zero at all other nodes. Specifically, we require that if x_j is the x-coordinate of node j, then

$$\phi_i(x_j) = \begin{cases} 1 & \text{if } i = j \\ 0 & \text{if } i \neq j \end{cases} \qquad (1.5.3)$$

In our example, $i = 1, 2, 3$ and $j = 0, 1, 2, 3, 4$. It is obvious that the piece-wise-linear basis functions shown in Fig. 1.3a satisfy (1.5.3) (e.g., $\phi_1(x_1) = 1$, but $\phi_1(x_j) = 0$ for $j = 0, 2, 3$, and 4), so that the basis functions defined in (1.5.1) satisfy criterion 3. Note that $i = 0, 4$ are not included since the basis functions are required to satisfy the homogeneous end conditions. We relax this later and introduce $\phi_0(x)$, $\phi_4(x)$ to treat more general boundary data in Chapter 2.

Let v_h denote a function in H_0^1. According to (1.4.7) and our new notation,

$$v_h(x) = \sum_{i=1}^{N} \beta_i \phi_i(x) \qquad (N = 3)$$

Let v_j denote the value of the function v_h at an arbitrary nodal point j (i.e., $v_j = v_h(x_j)$) and let (1.5.3) hold. For our example,

$$v_j = \sum_{i=1}^{3} \beta_i \phi_i(x_j) = \beta_j, \qquad j = 1, 2, 3$$

It follows that the finite element representation of v_h takes the form

$$v_h(x) = \sum_{i=1}^{N} v_i \phi_i(x), \qquad v_i = v_h(x_i) \qquad (1.5.4)$$

It is important to understand how the terms in (1.5.4) sum up to give a continuous representation of v_h. For example, suppose that the values of v_h at nodes 1, 2, and 3 are 0.9, 0.7, and 0.2, respectively. Then, substituting these values into (1.5.4) gives

$$v_h(x) = 0.9\phi_1(x) + 0.7\phi_2(x) + 0.2\phi_3(x)$$

so that these three components combine to give the continuous piecewise-linear function shown in Fig. 1.5.

There is a final point that should be made here which is suggested by the

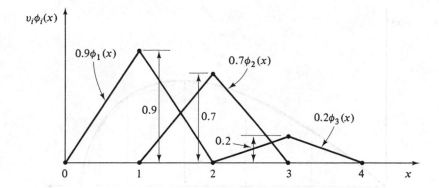

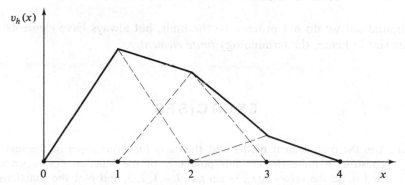

FIGURE 1.5   *Piecewise-linear finite element representation of the function* v_h.

form of the piecewise-linear function v_h in Fig. 1.5. Suppose that the actual solution to our model problem is the smooth function indicated in Fig. 1.6. If we consider the shape of this function over a sufficiently small subinterval of its domain, then it is clear that it is nearly linear over this interval, as indicated in the figure. If u is approximated by piecewise-linear functions with values coinciding with those of u at the nodes, the result is a polygonal function which closely resembles u. This is piecewise-linear *interpolation* of the exact solution u. As the mesh is *refined* (i.e., as the number of elements is increased and their size h is decreased), the finite element interpolant becomes progressively closer to u. On the other hand, our finite element approximation u_h to the solution of the boundary-value problem will also be piecewise-linear but its nodal values will generally not agree with those of the exact solution. In much the same way as the interpolant, the approximate solutions u_h would seem to have the property that they would produce progressively better approximations of u as the mesh is refined. These ideas, of course, are closely related to the concepts that underlie differential calculus. Naturally, in our

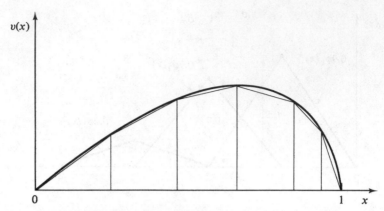

FIGURE 1.6 *Local linear interpolation of a smooth function* v.

computations we do not proceed to the limit, but always have elements of finite size h: hence, the terminology *finite elements*.

EXERCISES

1.5.1 Use the four-element model and the basis functions given in the text to construct a finite element interpolant v_h of the function $f(x) = \sin \pi x$; that is, set the values $v_h(x_i) = \sin \pi x_i$, $i = 1, 2, 3$, and plot the functions f and v_h. Also, plot the derivatives f' and v_h'.

1.5.2 Consider the use of piecewise-constant basis functions. Use elements with a single node, say at the center of the element. Discuss this choice of basis with respect to criteria 1, 2, and 3.

1.5.3 Consider a finite element with three nodes, one at each endpoint and one at the center of the element:

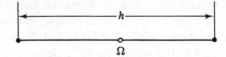

Since three values are sufficient to determine a parabola, it appears that quadratic polynomials might be used to construct basis functions with such elements:

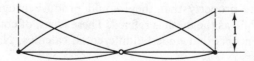

(a) Using these facts and the ideas discussed in the text, construct piecewise-quadratic basis functions $\phi_i(x)$ for a finite element mesh consisting of two such elements and five nodes. Sketch the $\phi_i(x)$ as functions on the domain $0 \leq x \leq 1$ and show that they satisfy the three criteria for finite element basis functions.

(b) Use this finite element model to construct an interpolant of the function $f(x) = \sin \pi x$; that is, construct v_h such that $v_h(x_j) = \sin \pi x_j$, x_j being a nodal-point coordinate. Plot the functions f and v_h.

1.5.4 It should be clear from Exercise 1.5.3 that polynomials of higher degree than one can be used to generate finite element basis functions.

(a) Describe how one might use cubic polynomials within each element to produce basis functions satisfying criteria 1, 2, and 3. How many nodes per element should there be? What would these functions look like?

(b) Repeat part (a) for quartic (fourth-degree) polynomials.

(c) Now consider the general case. Describe acceptable finite element basis functions for the case in which polynomials of degree n are used within each element.

1.6 FINITE ELEMENT CALCULATIONS

We return to the Galerkin approximation of the variational boundary-value problem (1.4.1) using finite element techniques to construct the basis functions ϕ_i. The approximate problem, then, consists of finding $u_h \in H_0^h$, where H_0^h is a subspace of H_0^1 defined by the particular choices of ϕ_i, such that

$$\int_0^1 (u_h' v_h' + u_h v_h)\, dx = \int_0^1 x v_h\, dx \qquad \text{for all } v_h \subset H_0^h \qquad (1.6.1)$$

in which $u_h = \sum_{i=1}^N u_i \phi_i$ and u_i are the values of u_h at the nodal points in the finite element mesh. In view of (1.4.14), this leads to the system of linear equations

$$\sum_{j=1}^N K_{ij} u_j = F_i, \qquad i = 1, 2, \ldots, N \qquad (1.6.2)$$

where K_{ij} and F_i are defined by (1.4.12) and (1.4.13).

We examine next some special and fundamentally important properties of the stiffness matrix **K** and the load vector **F**.

1. **Summability of Stiffnesses:** This is, perhaps the most important property of stiffness matrices computed using finite elements. Suppose that we use, as an example, the finite element mesh and basis functions indicated in Fig. 1.3a. In view of (1.4.12), each entry K_{ij} is obtained

by integrating $(\phi_i'\phi_j' + \phi_i\phi_j)$ over the entire domain $0 \leq x \leq 1$. But the operation of integration is additive (e.g., $\int_0^1 f\,dx = \int_0^{1/2} f\,dx + \int_{1/2}^1 f\,dx$), so that K_{ij} can be calculated as the sum

$$
\begin{aligned}
K_{ij} &= \int_0^1 (\phi_i'\phi_j' + \phi_i\phi_j)\,dx \\
&= \int_0^h (\phi_i'\phi_j' + \phi_i\phi_j)\,dx + \int_h^{2h} (\phi_i'\phi_j' + \phi_i\phi_j)\,dx \\
&\quad + \int_{2h}^{3h} (\phi_i'\phi_j' + \phi_i\phi_j)\,dx + \int_{3h}^1 (\phi_i'\phi_j' + \phi_i\phi_j)\,dx \\
&= \sum_{e=1}^4 \int_{\Omega_e} (\phi_i'\phi_j' + \phi_i\phi_j)\,dx
\end{aligned}
\tag{1.6.3}
$$

where $\int_{\Omega_e}$ denotes integration over element Ω_e.

Let the terms

$$
K_{ij}^e = \int_{\Omega_e} (\phi_i'\phi_j' + \phi_i\phi_j)\,dx
\tag{1.6.4}
$$

represent components of the *element stiffness matrix* for finite element Ω_e. Thus,

$$
K_{ij} = \sum_{e=1}^4 K_{ij}^e
\tag{1.6.5}
$$

Similarly,

$$
F_i = \sum_{e=1}^4 F_i^e, \qquad F_i^e = \int_{\Omega_e} x\phi_i\,dx
\tag{1.6.6}
$$

where F_i^e are components of the *element load vector* for finite element Ω_e.

The fact that K_{ij} and F_i can be computed as the sums of contributions from each element is a key feature of finite element methods. Because of this elementary property, *it is possible to generate* **K** *and* **F** *by computing only the element matrices* **K**^{e} *and* **F**^{e} *for a typical element* Ω_e *and then to construct* **K** *and* **F** *as the sums indicated in* (1.6.5) *and* (1.6.6). We elaborate on this fact later.

2. **Sparseness of K:**   For our model problem, with the mesh indicated in Fig. 1.3a, we must compute nine numbers, K_{ij}, $i,j = 1, 2, 3$, in order to arrive at the stiffness matrix for our approximation. However, an examination of Fig. 1.7 reveals that ϕ_1 and ϕ_1' are different

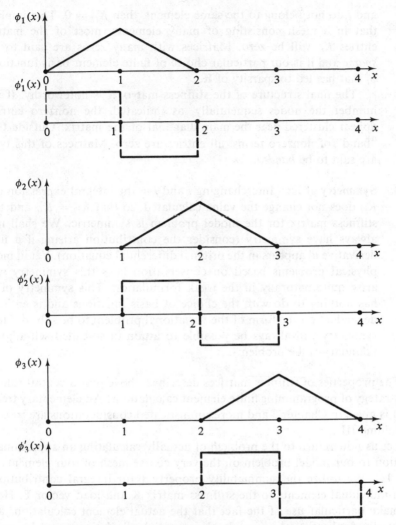

FIGURE 1.7 *Piecewise-linear basis functions and their derivatives.*

from zero only in elements Ω_1 and Ω_2 adjacent to node 1. Similarly, ϕ_2 and ϕ_2' are nonzero only on elements Ω_2 and Ω_3 adjacent to node 2, and ϕ_3 and ϕ_3' are nonzero only on elements Ω_3 and Ω_4 adjacent to node 3. Consequently, products $\phi_i\phi_j$ and $\phi_i'\phi_j'$ are nonzero only where the supports for basis functions ϕ_i and ϕ_j "overlap." For example, the products $\phi_1\phi_2$ and $\phi_1'\phi_2'$ are nonzero only on element 2, whereas $\phi_1\phi_3$ and $\phi_1'\phi_3'$ are zero everywhere. Hence, the integrals K_{12}, K_{21} are nonzero but $K_{13} = K_{31} = 0$ automatically. It follows that if nodes i

and j do not belong to the same element, then $K_{ij} = 0$. This implies that in a mesh consisting of many elements, most of the matrix entries K_{ij} will be zero. Matrices with many zeros are said to be *sparse* and it is our particular choice of finite element basis functions ϕ_i that has led to sparsity of $\mathbf{K}$.

The final structure of the stiffness matrix $\mathbf{K}$ is noteworthy. If we number the nodes sequentially, as indicated, the nonzero entries appear clustered near the main diagonal of the matrix. Outside this "band" of nonzero terms, all entries are zero. Matrices of this type are said to be *banded*.

3. **Symmetry of K:**   Interchanging i and j in the integral expression for K_{ij} does not change the value calculated, so that $K_{ij} = K_{ji}$ and the stiffness matrix for the model problem is symmetric. We shall not always have symmetry (consider the contribution arising if a first derivative u' appears in the original differential equation). Yet in most physical problems based on conservation laws this symmetry will arise quite naturally in the weak formulation. This symmetry of $\mathbf{K}$ has nothing to do with the choice of basis functions and is entirely dependent on the form of the variational problem to be solved. Such symmetry will always be possible to attain in so-called self-adjoint boundary-value problems.

The properties of stiffness matrices described above play a central role in the strategy of programming finite element calculations. An elementary treatment is given in Chapter 3 and more sophisticated considerations are treated in Volume III.

Let us now return to the problem of actually calculating an approximate solution to our model problem on the very coarse mesh of four elements in Fig. 1.2. We utilize the summability property 1 for integral contributions from individual elements to the stiffness matrix $\mathbf{K}$ and load vector $\mathbf{F}$. Here we make particular use of the fact that the actual element calculations are essentially duplicative. As a result, the essential calculations need be made only on a single typical finite element Ω_e.

We begin the calculation of element matrices by considering the representative element Ω_e shown in Fig. 1.8 and introducing the local node indices A and B. Let ξ be a local coordinate on this representative element with its origin at the left node A of Ω_e. Then as x traverses x_A to x_B, ξ goes from 0 to h. We have, simply, $\xi = x - x_A$.

Recall that the basis functions ϕ_i are constructed by piecing together polynomials defined locally over each element (Fig. 1.3b). We refer to these component parts as *element shape functions*. For example, the basis function

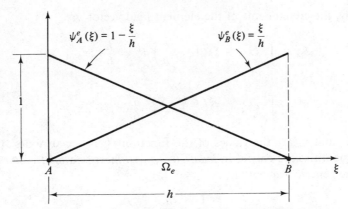

FIGURE 1.8 *A generic finite element Ω_e with linear shape functions.*

ϕ_A at node A in the mesh is produced by combining element shape functions defined on the elements connected at node A.

Let ψ_A^e and ψ_B^e denote the shape functions defined for element Ω_e and shown in Fig. 1.8. Since these are simply parts of ϕ_A and ϕ_B, these shape functions are given in terms of the local coordinate ξ by

$$\psi_A^e(\xi) = 1 - \frac{\xi}{h}, \qquad \psi_B^e(\xi) = \frac{\xi}{h}$$

Clearly,

$$\psi_A^{e\prime}(\xi) = -\frac{1}{h}, \qquad \psi_B^{e\prime}(\xi) = \frac{1}{h}$$

According to (1.6.4), the element matrix coefficients for our generic element Ω_e are

$$k_{AA}^e = \int_0^h \{[\psi_A^{e\prime}(\xi)]^2 + [\psi_A^e(\xi)]^2\}\, d\xi$$

$$= \int_0^h \left[\frac{1}{h^2} + \left(1 - \frac{\xi}{h}\right)^2\right] d\xi = \frac{1}{h} + \frac{h}{3}$$

$$k_{AB}^e = k_{BA}^e = \int_0^h [\psi_A^{e\prime}(\xi)\psi_B^{e\prime}(\xi) + \psi_A^e(\xi)\psi_B^e(\xi)]\, d\xi$$

$$= \int_0^h \left[\left(\frac{-1}{h}\right)\frac{1}{h} + \left(1 - \frac{\xi}{h}\right)\frac{\xi}{h}\right] d\xi = -\frac{1}{h} + \frac{h}{6}$$

and

$$k_{BB}^e = \int_0^h \{[\psi_B^{e\prime}(\xi)]^2 + [\psi_B^e(\xi)]^2\}\, d\xi = \frac{1}{h} + \frac{h}{3}$$

Similarly, the components of the element load vector are

$$F_A^e = \int_0^h (x_A + \xi)\left(1 - \frac{\xi}{h}\right) d\xi = \frac{h}{6}(2x_A + x_B)$$

and

$$F_B^e = \int_0^h (x_A + \xi)\left(\frac{\xi}{h}\right) d\xi = \frac{h}{6}(x_A + 2x_B)$$

where x_A and x_B are the values of the function $f(x) = x$ at nodes A and B. These quantities are entries in *local element stiffness* and *load vectors* $\mathbf{k}^e$ and $\mathbf{f}^e$ for the generic element Ω_e:

$$\mathbf{k}^e = \begin{bmatrix} \dfrac{1}{h} + \dfrac{h}{3} & -\dfrac{1}{h} + \dfrac{h}{6} \\ -\dfrac{1}{h} + \dfrac{h}{6} & \dfrac{1}{h} + \dfrac{h}{3} \end{bmatrix}, \qquad \mathbf{f}^e = \frac{h}{6}\begin{bmatrix} 2x_A + x_B \\ x_A + 2x_B \end{bmatrix} \qquad (1.6.7)$$

It is these element matrices that are actually calculated in a finite element computer code. When the dimension of the problem is specified (in this example, there will be only three equations) and when the coordinates of nodes in each element are specified, the entries in (1.6.7) are computed and stored in the row i and column j appropriate for the nodes and elements they represent. In this way, the expanded element matrices $\mathbf{K}^e$ and $\mathbf{F}^e$ are then essentially generated by the summations (1.6.5) and (1.6.6). Since $h = \frac{1}{4}$ in the present example, the use of (1.6.7) and the process just described lead to the following expanded element matrices:*

Element Ω_1

$$\mathbf{K}^1 = [K_{ij}^1] = \frac{1}{24}\begin{bmatrix} 98 & 0 & 0 \\ 0 & 0 & 0 \\ 0 & 0 & 0 \end{bmatrix}, \qquad \mathbf{F}^1 = \{F_i^1\} = \frac{1}{96}\begin{bmatrix} 2 \\ 0 \\ 0 \end{bmatrix}$$

Element Ω_2

$$\mathbf{K}^2 = [K_{ij}^2] = \frac{1}{24}\begin{bmatrix} 98 & -95 & 0 \\ -95 & 98 & 0 \\ 0 & 0 & 0 \end{bmatrix}, \qquad \mathbf{F}^2 = \{F_i^2\} = \frac{1}{96}\begin{bmatrix} 4 \\ 5 \\ 0 \end{bmatrix}$$

---

* One somewhat artificial feature of the present example which is not typical of usual finite element calculations is the manner in which boundary conditions enter the formulation. We have chosen to follow the present tack to emphasize the connection between the finite element approximation and Galerkin's approximation described earlier. However, a remarkable and extremely useful property of finite element formulations is that the boundary conditions can be quite general and need be specified only at a later stage of the analysis. We take up this subject in greater detail in Chapter 2.

Element Ω_3

$$\mathbf{K}^3 = [K^3_{ij}] = \frac{1}{24}\begin{bmatrix} 0 & 0 & 0 \\ 0 & 98 & -95 \\ 0 & -95 & 98 \end{bmatrix}, \qquad \mathbf{F}^3 = \{F^3_i\} = \frac{1}{96}\begin{bmatrix} 0 \\ 7 \\ 8 \end{bmatrix}$$

Element Ω_4

$$\mathbf{K}^4 = [K^4_{ij}] = \frac{1}{24}\begin{bmatrix} 0 & 0 & 0 \\ 0 & 0 & 0 \\ 0 & 0 & 98 \end{bmatrix}, \qquad \mathbf{F}^4 = \{F^4_i\} = \frac{1}{96}\begin{bmatrix} 0 \\ 0 \\ 10 \end{bmatrix}$$

Thus, according to (1.6.5) and (1.6.6),

$$\mathbf{K} = [K_{ij}] = \mathbf{K}^1 + \mathbf{K}^2 + \mathbf{K}^3 + \mathbf{K}^4 = \frac{1}{24}\begin{bmatrix} 196 & -95 & 0 \\ -95 & 196 & -95 \\ 0 & -95 & 196 \end{bmatrix}$$

$$\mathbf{F} = \{F_i\} = \mathbf{F}^1 + \mathbf{F}^2 + \mathbf{F}^3 + \mathbf{F}^4 = \frac{1}{96}\begin{bmatrix} 6 \\ 12 \\ 18 \end{bmatrix}$$

and our final system of equations is

$$\frac{1}{24}\begin{bmatrix} 196 & -95 & 0 \\ -95 & 196 & -95 \\ 0 & -95 & 196 \end{bmatrix}\begin{bmatrix} u_1 \\ u_2 \\ u_3 \end{bmatrix} = \frac{1}{16}\begin{bmatrix} 1 \\ 2 \\ 3 \end{bmatrix} \qquad (1.6.8)$$

where, again, u_1, u_2, and u_3 are the values of u_h at nodes 1, 2, and 3, respectively. The procedure used to obtain the stiffness matrix in (1.6.8) can be described by the flowchart in Fig. 1.9.

Upon solving (1.6.8), we find that, to four places,

$$\mathbf{u} = \begin{bmatrix} u_1 \\ u_2 \\ u_3 \end{bmatrix} = \begin{bmatrix} 0.0353 \\ 0.0569 \\ 0.0505 \end{bmatrix}$$

Thus, by (1.5.4), the finite element approximation u_h of the solution to (1.2.1) is

$$u_h(x) = 0.0353\phi_1(x) + 0.0569\phi_2(x) + 0.0505\phi_3(x)$$

where the ϕ_i are the functions shown in Fig. 1.7. We discuss properties of this approximate solution in the next section.

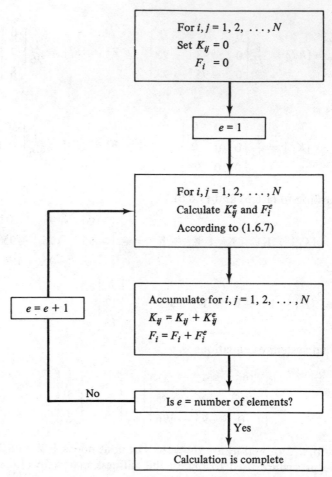

FIGURE 1.9 *Flow chart showing element by element calculation and assembly of **K** and **F**.*

EXERCISES

1.6.1 Calculate the load vector **F** for a finite element model consisting of four elements of equal length and piecewise-linear basis functions for the following boundary-value problem:

$$-u'' + u = 1, \qquad 0 < x < 1$$
$$u(0) = u(1) = 0$$

1.6.2 Suppose that 10 equal elements are used to approximate the model problem (1.2.1). Calculate all of the nonzero contributions to the stiffness matrix **K** from the fourth element (element Ω_4 is the subinterval $0.3 \leq x \leq 0.4$).

1.6.3 Using the four-element model shown in Fig. 1.3a, calculate matrix **K** and the vector **F** for the boundary-value problem

$$-u'' = 1, \qquad 0 < x < 1$$
$$u(0) = u(1) = 0$$

Solve for **u** and plot u and u_h as functions of x.

1.7 INTERPRETATION OF THE APPROXIMATE SOLUTION

In applications of the finite element method, the analysis is only partially completed by the calculation of the vector of nodal values **u**. There remains the very important task of the interpretation of the approximate solution that has been found. The answers to qualitative questions, such as "What is the general character of the solution?" or "Where are the regions in which the solution varies most rapidly?" are usually of interest, and these are often best answered by examining graphs of the solution and its derivative. Plots of the finite element solution not only reveal qualitative features of the solution but also provide an easy test for detecting modeling or data errors.

The finite element approximation u_h to the solution of the model problem described earlier and its derivative u_h' are shown in Fig. 1.10. The following observations are made:

1. The approximate solution is a fairly smooth function—there are no apparent oscillations or regions of very high gradients.

2. There is one extremum of the solution, a maximum value of 0.0569 located at $x = 0.5$.

3. The derivative of the solution is large near the endpoints, the largest absolute value occurring near $x = 1.0$.

Of course, if we use a "finer" finite element mesh (i.e., more elements of smaller length), then we could sharpen our picture of these features of the solution, but the rather crude approximation we have calculated suits our present purposes.

To see how these features have relevance in a physical problem, let us suppose that the model problem has arisen in the analysis of a stretched string, supported on an elastic foundation and subjected to a transverse load whose distribution is given by $f(x) = x$. The solution u is the transverse deflection of the string and its derivative u' is proportional to the stress in

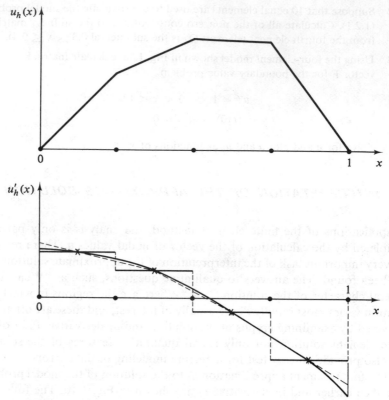

FIGURE 1.10 *A finite element approximation of the solution of the model problem and its derivative.*

the string. Moreover, the total strain energy in the system (i.e., in the string and the elastic support) is given by

$$U = \tfrac{1}{2} \int_0^1 [(u')^2 + u^2]\, dx \tag{1.7.1}$$

In such an application, we might seek answers to the following questions:

1. What is the location and the value of the maximum deflection?
2. What is the location and the value of the maximum stress?
3. What is the value of total strain energy in the system?

The best answer to question 1 that we can extract from our approximation is that u_h has a peak at node 2, and we conclude from this that the maximum deflection occurs at $x = 0.5$ and is given by $u_2 = 0.0569$, as noted earlier. This is slightly in error, but is the best information available from our crude approximation. Figure 1.10 indicates that the answer to question 2 is not so straightforward. It is clear that the maximum value of $|u'_h|$ is 0.202 and that this value occurs throughout element Ω_4. On both physical and mathematical grounds we know that the stress in the model problem varies continuously. To which point in element Ω_4, then, shall we assign the calculated value of maximum stress? One obvious answer is to consider the calculated value to be the average value over the element and to assign it to the midpoint of the element (i.e., the point $x = 0.875$). This choice, as it turns out, is the best one we can make, but it falls short of a satisfactory answer to our original question.

Figure 1.10 shows that we should expect the maximum stress to occur at the point $x = 1$. In order to infer, from the finite element solution, the value of the derivative at the endpoint $x = 1$, we might plot, as in Fig. 1.10. the values of the "stress" u'_h at the element midpoints. We could *extrapolate* the values of stress to the boundary point, $x = 1$, by sketching a smooth curve through these points as indicated by the solid curve in the figure, but this procedure is too imprecise to be reliable and is not general. Alternatively, we could fit a curve, say a straight line, to midpoint values of u'_h to evaluate this function at points of interest. The dashed line in the figure illustrates this procedure. Extensions of this procedure to multidimensional problems and automation of it in finite element programs are possible and, in fact, not uncommon.

The strain energy of the finite element solution is easily evaluated. The integral in (1.7.1) can be calculated element by element, as in the evaluation of the stiffness coefficients, K_{ij}, and the results summed over the elements. It is not difficult to see, however, that the calculations required have already been carried out in our earlier computations.

Let $\boldsymbol{\phi}$ denote the vector of basis functions and $\mathbf{u}$ the vector of nodal values of the solution,

$$\boldsymbol{\phi} = \begin{bmatrix} \phi_1 \\ \phi_2 \\ \phi_2 \end{bmatrix}, \qquad \mathbf{u} = \begin{bmatrix} u_1 \\ u_2 \\ u_3 \end{bmatrix}$$

Then,

$$u_h(x) = \mathbf{u}^T \boldsymbol{\phi}(x)$$

where $(\cdot)^T$ denotes transposition, and the approximate energy is

$$U_h = \tfrac{1}{2} \int_0^1 [(u'_h)^2 + u_h^2] \, dx$$

$$= \tfrac{1}{2} \mathbf{u}^T \int_0^1 (\boldsymbol{\phi}'^T \boldsymbol{\phi}' + \boldsymbol{\phi}^T \boldsymbol{\phi}) \, dx \, \mathbf{u}$$

$$= \tfrac{1}{2} \mathbf{u}^T \mathbf{K} \mathbf{u}$$

But $\mathbf{Ku} = \mathbf{F}$, so

$$U_h = \tfrac{1}{2}\mathbf{u}^T\mathbf{F} \tag{1.7.2}$$

Carrying out the calculation indicated by (1.7.2) gives the value of energy in the finite element solution of $U_h = 0.0094$.

Calculations and observations such as these indicate that a great deal of useful information can be extracted from a careful examination of properties of the solution.

EXERCISE

1.7.1 For piecewise-linear finite element approximation of the model problem in which six equal elements are used, the nodal point values of $u_h(x)$, are

$$\mathbf{u}^T = [0, 0.0242, 0.0445, 0.0567, 0.0565, 0.0394, 0]$$

and for an eight-element model, they are

$$\mathbf{u}^T = [0, 0.0184, 0.0351, 0.0484, 0.0567, 0.0579, 0.0503, 0.0318, 0]$$

(a) Plot u_h (on the same graph) for the four-, six-, and eight-element solutions. Note the location and value of maximum deflection.

(b) Plot curves through the element values of u'_h for the three solutions. Extrapolate to the boundary to find the maximum stress predicted by each solution.

(c) Based on the data available, give your best estimate of the location and value of maximum deflection and stress in the model problem. Compare your estimates with the exact values.

(d) Calculate the energy U_h for the six-element mesh and compare it with the exact energy of the solution of the model problem.

1.8 ACCURACY OF THE FINITE ELEMENT APPROXIMATION

Since we have taken for granted that our mission is to compute approximations of solutions of boundary-value problems, the question of the accuracy of our approximation naturally arises. Indeed, any responsible user of finite element methods should be concerned about the question of accuracy and with the issue of how the error in the approximate solutions is affected as the number of elements in the mesh is increased.

The *error* in the finite element approximation is the function e, defined as the difference between the exact and the approximate solutions,

$$e(x) = u(x) - u_h(x) \tag{1.8.1}$$

It is clear that the actual error can never be calculated unless the exact solution is known, so that initially there may seem to be little reason to want to compute it. However, even when u is unknown, it is possible to construct estimates of the error and to determine if the error decreases as h decreases and the number of elements becomes larger. Information of this type is of great utility in finite element calculations; it can be used to judge the acceptability of various choices of elements and to estimate the increase in accuracy that one might expect if the number of elements were doubled or tripled. More important, such estimates of the error would also indicate which elements were unacceptable for use in the problem at hand.

Since errors are functions, we must have some means of measuring the size of functions if we are to discuss the accuracy of our solutions. A natural and universally used measure of the magnitude of a function g is a nonnegative number, associated with g, called the *norm* of g and written $\|g\|$. If $g \equiv 0$, then $\|g\| = 0$; conversely, if $\|g\| = 0$, then g is interpreted as zero. Thus, we wish to calculate or estimate the error when measured with respect to some appropriate norm.

There are three principal choices of error norms that are commonly used in finite element methods: the energy norm $\|e\|_E$, the mean-square norm $\|e\|_0$, and the maximum or infinity norm $\|e\|_\infty$.

The form of the energy norm is dependent upon the form of the boundary-value problem under consideration. For our model problem, $\|\cdot\|_E$ can be taken to be the square root of twice the energy U given by (1.7.1), so that

$$\|e\|_E = \left\{ \int_0^1 [(e')^2 + e^2]\, dx \right\}^{1/2} \tag{1.8.2}$$

The energy norm is actually one of the most natural and meaningful ways to quantify the error in our approximation. If we demand that our approximation be such that the error is well behaved with respect to the energy norm as the mesh is refined, then we can generally be confident that we have an acceptable method of approximation.

The mean-square (L^2 or "L-two") norm measures the root-mean-square of a function over its domain and is defined by

$$\|e\|_0 = \left(\int_0^1 e^2\, dx \right)^{1/2} \tag{1.8.3}$$

whereas the maximum norm measures the maximum absolute value of a function over its domain:

$$\|e\|_\infty = \max_{0 \le x \le 1} |e(x)| \tag{1.8.4}$$

It is customary to seek "asymptotic estimates" of these error measures in finite element approximations. By this, we mean the following: suppose that the domain of our problem is discretized by a finite element mesh consisting of elements of equal length h. We *refine* this mesh by decreasing h and, of course, increasing the number of elements in the mesh. We wish to know bounds on the error, measured in the foregoing norms, which are valid for h very small and which are given in terms of the *mesh parameter h*. Ordinarily, these estimates will be of the form

$$\|e\| \leq Ch^p \tag{1.8.5}$$

where C is a constant depending upon the data of the problem and p is an integer that depends upon the basis functions chosen in our finite element approximation. The exponent p is a measure of the *rate of convergence* of the method with respect to the particular choice of the norm $\|\cdot\|$. If p is positive, then ($\|e\|$ being nonnegative) the error $\|e\|$ clearly approaches zero as h tends to zero. When $\|e\|$ approaches zero, we say that our approximation *converges* to the exact solution *with respect to the norm* $\|\cdot\|$. It is possible that we might have situations in which convergence is obtained with respect to one norm but not with respect to another; in fact, the notion of convergence is unquestionably norm-dependent.

Since estimates such as (1.8.5) require no information from the actual finite element solution, they are known prior to the construction of the solution and are called *a-priori* estimates. In some cases more detailed estimates of accuracy can be based on information obtained from the finite element solution itself. Such estimates, called *a-posteriori* estimates, can be calculated only after the finite element solution has been calculated.

In the case of our model problem, the following a-priori error estimates can be shown to hold for piecewise-linear basis functions:*

$$\left.\begin{array}{l} \|e\|_E \leq C_1 h \\ \|e\|_0 \leq C_2 h^2 \\ \|e\|_\infty \leq C_3 h^2 \end{array}\right\} \tag{1.8.6}$$

There are, of course, other measures of error that are of practical interest. We could, for instance, test

$$\|e'\|_\infty = \max_{0 \leq x \leq 1} |e'(x)|$$

or, in some cases

$$\|e''\|_\infty, \quad \|e'\|_0, \quad \text{or} \quad \|e''\|_0$$

or we could check the pointwise behavior of the error by calculating $|e(\xi)|$, where ξ is a given point in the domain of the solution. In regard to such pointwise tests, it is a remarkable fact that many finite element methods exhibit what is called *superconvergence* when special points ξ are selected. By

---

* We study such estimates in Volume IV of this series.

superconvergence it is meant that extraordinarily high rates of convergence (and concomitant high accuracy) are observed at these points. The error measures (1.8.6), however, are generally considered to be the most natural and useful choices in finite element calculations.

Returning to (1.8.5), we note that the graph of a real-valued function $E(h)$ of the form Ch^p is a straight line of slope p and intercept $\log C$ on a log-log plot:

$$\log E = p \log h + \log C \qquad (1.8.7)$$

Thus, if the equality in (1.8.5) holds, a plot of $\log \|e\|$ versus $\log h$ will produce a straight line, for h sufficiently small, the slope of which is the rate of convergence with respect to the norm $\| \cdot \|$; the constant C will shift the line up or down. Figure 1.11 contains such error plots for our model problem for errors

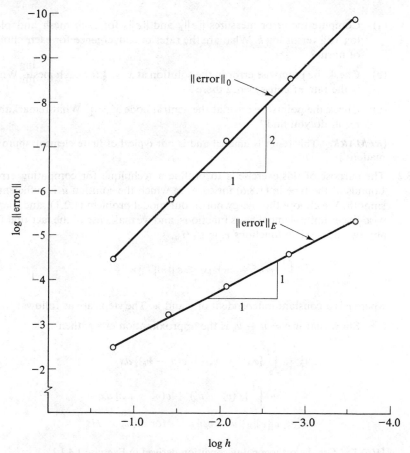

FIGURE 1.11 *Log-log plots of the errors* $\|e\|_E$, $\|e\|_0$ *as a function of* h *for the model problem.*

in the energy and mean-square norms. In the calculation of these results, a uniform mesh (i.e., a mesh with all elements of equal length h) and piecewise-linear basis functions of the type described earlier were used. We observe that these computed results confirm the a priori estimates given in (1.8.6).

EXERCISES

1.8.1 Compute the exact solution and finite element approximations of it for two, four, and six elements of the boundary-value problem

$$-u'' = 1 - x^2, \qquad 0 < x < 1$$
$$u(0) = 0, \qquad u(1) = 0$$

(a) Compute the error measures $\|e\|_E$ and $\|e\|_0$ for each mesh and plot $\log \|e\|$ versus $\log h$. What are the rates of convergence for each choice of norm?

(b) Check the pointwise error in the solution at $x = \frac{1}{8}$ for each mesh. What is the rate of convergence there?

(c) Check the pointwise error at the central node $x = \frac{1}{2}$. What remarkable result do you find?

[*REMARK:* This result is unusual and is not typical of finite element approximations.]

1.8.2 The purpose of this exercise is to outline a technique for computing error bounds of the type in (1.8.6) for cases in which the solution u is sufficiently smooth. We choose the energy norm, our model problem (1.2.1), and piecewise-linear finite element basis functions, and we make use of the fact that for any two admissible functions $v, w \in H_0^1$,

$$\int_0^1 (v'w' + vw) \, dx \le c \, \|v\|_E \, \|w\|_E$$

where c is a constant independent of v and w. The steps are as follows:

(a) Show that if $e = u - u_h$ is the approximation error, then

$$\|e\|_E^2 = \int_0^1 [e'(u' - v_h') + e(u - v_h)] \, dx$$
$$+ \int_0^1 [e'(v_h' - u_h') + e(v_h - u_h)] \, dx$$
$$\le c\|e\|_E \|u - v_h\|_E \qquad \text{for all } v_h \in H_0^h$$

[*HINT:* Use the orthogonality condition derived in Exercise 1.4.1.]

(b)   Let $w_h \in H_0^h$ be that particular finite element test function which inter-polates u; that is, w_h coincides exactly with u at the nodes. Show that

$$\| u - w_h \|_E^2 \le c_1 [\max_{0 \le x \le 1} |u'(x) - w_h'(x)|^2 + \max_{0 \le x \le 1} |u(x) - w_h(x)|^2]$$

$$\le c_2 h^2 \quad \text{(as } h \text{ approaches zero)}$$

where c_1 and c_2 are constants.

[*HINT*:   For this purpose, assume that u has continuous second derivatives and show that on a typical element,

$$|u(\xi) - w_h(\xi)| = |u'(x_A)\xi - \frac{w_B - w_A}{h}\xi + \frac{1}{2}u''(\zeta)\xi^2| \le c_3 h^2$$

where $0 \le \xi \le h$, x_A and x_B are coordinates of the endpoint nodes, ζ is a point within the element, and c_3 is a constant. Then derive a similar result for $|u'(\xi) - w_h'(\xi)|$.]

(c)   Since the result in part (a) holds for any v_h in H_0^h, pick $v_h = w_h$ and show that

$$\| e \|_E \le Ch$$

for some constant C and h sufficiently small.

[*REMARK*:   Although the method for constructing error estimates in this exercise cannot be used for problems in which the solution is not smooth (such as problem (1.2.2)), it is possible to obtain estimates of the type in (1.8.6) for such problems by using more powerful methods of analysis. We discuss these issues in Volume IV.]

2

ONE-DIMENSIONAL PROBLEMS

2.1 INTRODUCTION

The great popularity of finite element methods is due to their success in solving complex two- and three-dimensional boundary-value problems and not in treating simple one-dimensional cases of the type considered in Chapter 1. Nevertheless, a great deal can be learned about the basic principles underlying finite elements by considering one-dimensional problems.

With this fact in mind, we continue our study of some one-dimensional boundary-value problems in this chapter with the objective of expanding and clarifying certain concepts which previously were dealt with only superficially. In particular, we consider a quite general class of second-order linear boundary-value problems with general boundary conditions, and we describe how acceptable finite element approximations of these problems can be constructed.

2.2 CLASSICAL STATEMENTS OF SECOND-ORDER, TWO-POINT BOUNDARY-VALUE PROBLEMS

2.2.1 Linear Second-Order Problems

We recall from our study of the model problem that a boundary-value problem is characterized by a differential equation, which must be satisfied by the solution on the interior Ω of its domain, and by boundary conditions that the solution must satisfy at the endpoints of Ω (in our model problem,

Ω was the interval $0 < x < 1$). A more general situation is governed by a differential equation of the form

$$a_0(x) \frac{d^2u(x)}{dx^2} + a_1(x) \frac{du(x)}{dx} + a_2(x)u(x) = f(x) \qquad (2.2.1)$$

wherein the coefficients a_0, a_1, and a_2 as well as the function f are given functions of x defined on the given domain.

This is a linear ordinary differential equation of second order in the function $u = u(x)$. The differential equation (2.2.1) is *linear* because the derivatives of the solution appear linearly in the equation. As a consequence of linearity, the "principle of superposition" applies: if u_1 is a solution of (2.2.1) for a choice of the right-hand side of $f = f_1$ and if u_2 is a solution corresponding to $f = f_2$, then, for any arbitrary constants α and β, $\alpha u_1 + \beta u_2$ is a solution satisfying (2.2.1) for the choice $f = \alpha f_1 + \beta f_2$. Equation (2.2.1) is of *second-order* because u'' is the derivative of highest order in (2.2.1).

We shall actually consider a situation in which the differential equation (2.2.1) holds only in certain subintervals of Ω and in which special conditions are to be met at the endpoints of these subintervals. We elaborate on these details later in this section.

It is well known that when we "integrate" (2.2.1) on an interval we will obtain a general solution that contains exactly two arbitrary constants. The complete specification of the solution must include, in addition to the differential equation, enough additional conditions to determine the values of these constants. These conditions take the form of conditions at the boundaries of the interval.

Throughout this chapter and, indeed, throughout most of this book, we shall be concerned with *elliptic* boundary-value problems in which the leading coefficient $a_0(x)$ neither changes sign nor vanishes (i.e., $|a_0(x)| > \gamma > 0$ for all x in the interval), and in which exactly half of the boundary conditions are specified at each endpoint. Such one-dimensional problems are called *two-point boundary-value problems*. Of course, it is perfectly all right to consider problems in which both conditions are applied at, say, $x = 0$, but then the problem takes on the form of an initial-value problem in which a solution is advanced in "time" $t = x$ from a set of initial conditions. We shall take up such time-dependent problems briefly in Chapter 6, but for now, the elliptic two-point boundary-value problem is of central importance. An understanding of properties of elliptic problems and their approximations is essential for a detailed study of other types of boundary-value and initial-value problems.

Second derivatives of the solution are specified on the interior of the interval by the differential equation itself; boundary conditions should involve only the values of the solution u and of its first derivatives. Thus, if (2.2.1) holds on the interval $0 < x < l$, the most general boundary conditions that satisfy the conventions we have established and that preserve the linearity of the boundary-value problem are of the form

$$\alpha_0 \frac{du(0)}{dx} + \beta_0 u(0) = \gamma_0$$

$$\alpha_l \frac{du(l)}{dx} + \beta_l u(l) = \gamma_l$$

(2.2.2)

where α_0, β_0, γ_0, α_l, β_l, and γ_l are given constants.

In summary, a linear, elliptic, second-order boundary-value problem is specified by the differential equation (2.2.1) and the boundary conditions (2.2.2). The data in the problem consist of the domain $0 < x < l$; the functions a_0, a_1, a_2, and f given on the interval $0 < x < l$; and the numbers α_0, β_0, γ_0, α_l, β_l, and γ_l.

In actual problems arising in models of physical systems, (2.2.1) and (2.2.2) may only partially characterize a two-point problem; important additional considerations that must be made to handle possible discontinuities in the data are discussed in the following subsection.

2.2.2 Physical Origins of Two-Point Problems

We now enter a rather detailed discussion of mathematical models of a class of physical phenomena that lead to two-point boundary-value problems. Our goal here is not only to demonstrate how such problems arise in physics but also to point out how discontinuities in data arise in physical problems and, later, to illustrate in a more general setting how these discontinuities are dealt with naturally in variational statements of such problems.

The mathematical description of time-independent physical phenomena often leads to elliptic boundary-value problems of the type described above. We now sketch the development of such problems in an abstract setting (i.e., without reference to a specific physical situation). We shall show that both the differential equation and the additional conditions necessary for a complete specification of the problem can be obtained from the smoothness conditions required of the solution and a physical principle which we refer to as a *conservation law*. Our results will be applicable to problems drawn from a wide variety of technical areas; applications to specific classes of physical problems can be obtained simply by identifying the appropriate physical variables and principles for the application at hand.

Most physical problems are formulated in terms of two functions whose values are to be determined. We refer to these functions as the *state variable*

u and the *flux* σ.* These quantities are related to each other by a *constitutive equation*, which contains all of the information descriptive of the particular material in which the process is taking place. In various classical theories of mathematical physics the constitutive equations are given the status of "laws" of physics, but they generally have the same mathematical form.

The form of the constitutive equation describing linear material behavior is

$$\sigma(x) = -k(x)\frac{du(x)}{dx} \qquad (2.2.3)$$

The function k will be referred to as the *material modulus* and is part of the data given in our boundary-value problem. Clearly, the constitutive equation (2.2.3) indicates that any nonuniformity in the state variable u will produce a flux σ. The flux produced at any point x in the physical environment in which this phenomenon is taking place will, according to (2.2.3), be proportional to the rate at which u changes; the proportionality is determined by k, which may vary from point to point and which may be different for different materials. We will take for granted that $k(x)$ is always either strictly positive or strictly negative (i.e., it will never be zero in the cases we wish to consider).

A *conservation law* is a statement involving the flux which requires that in every subdomain of the problem, the net flux entering the subdomain is zero (i.e., the flux is conserved). Flux can enter a part of the body in two ways: it can be supplied by a distribution of internal sources, denoted by the given function f, and it can enter through the boundary of the region. The table in Fig. 2.1 indicates the interpretation of u, σ, k, and f for various types of physical problems.

In addition to the conservation law and constitutive equation, certain other conditions are imposed directly on the state variable u. For instance, in all problems of interest, u is required to be a continuous function of x. In addition to this continuity requirement, which is essential in all problems, we frequently require that the state variable assume a given value on either (or both) boundaries of the body. If the particular physical problem includes this requirement, this condition is called an *essential* boundary condition. All other conditions in the boundary-value problem are derivable from the conservation principle.

To fix ideas, let us now consider a representative one-dimensional model of a physical system in which the conventions just described are in force. Consider, for instance, a one-dimensional body (e.g., a rod) occupying the region $0 < x < l$, as shown in Fig. 2.2a. The body is composed of two materials, one occupying the region $0 < x < x_1$ and the other the region

---

* The choice of names in the abstract setting is somewhat arbitrary. Another frequently used choice denotes u as the *displacement* and σ as the *stress*.

Physical Problem	Conservation Principle	State Variable, u	Flux, σ	Material Modulus, k	Source, f	Constitutive Equation, $\sigma = -ku'$
Deformation of an elastic bar	Equilibrium of forces (conservation of linear momentum)	Displacement	Stress	Young's modulus of elasticity	Body forces	Hooke's law
Heat conduction in a rod	Conservation of energy	Temperature	Heat flux	Thermal conductivity	Heat sources	Fourier's law
Fluid flow	Conservation of linear momentum	Velocity	Shear stress	Viscosity	Body forces	Stokes' law
Electrostatics	Conservation of electric flux	Electric potential	Electric flux	Dialectric permittivity	Charge	Coulomb's law
Flow through porous media	Conservation of mass	Hydraulic head	Flow rate	Permeability	Fluid source	Darcy's law

FIGURE 2.1 Interpretation of physical variables and equations for various types of physical problems.

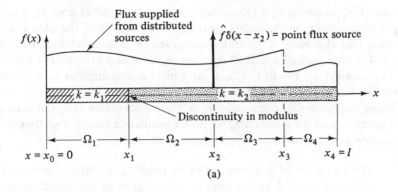

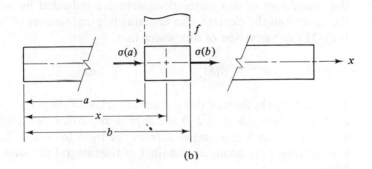

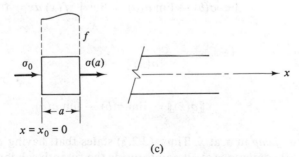

FIGURE 2.2 *(a) a one-dimensional body naturally subdivided into four regions Ω_i, i = 1, 2, 3, 4, between discontinuities in the material modulus k and the source f; (b) fluxes on an interior element, and (c) fluxes on an element containing a boundary point.*

$x_1 < x < l$. At point x_1, the material modulus k suffers a simple discontinuity, as indicated, but is continuous within each of the two material sections. In this example, the distribution f of internal sources is assumed to be continuous at all points except at $x = x_2$, where a *concentrated* source of intensity $\hat{f}$, represented by a Dirac delta, is imposed, and at $x = x_3$, where a simple discontinuity in the source distribution exists. Thus, the data k and f are such that the body is naturally composed of four subdomains Ω_i, $i = 1, 2, 3, 4$, within each of which all data are smooth, and five points (including the boundaries) x_i, $i = 0, 1, 2, 3, 4$, at which discontinuities in some data occur. We shall refer to the Ω_i as *smooth subdomains*.

Using only essential conditions on u, the conservation law, the constitutive equation, and the given data, we now formulate a classical mathematical description of this one-dimensional physical problem.

1. The flux must be conserved at every point in the body. Consider first a typical point $\bar{x}$ in the interior of a region of smooth data and a region $a \le x \le b$ containing $\bar{x}$, as shown in Fig. 2.2b. The fluxes on the boundaries of this material element are indicated by arrows in the figure and the element also contains internal sources of intensity $f(x)$. The conservation of flux states that

$$\sigma(b) - \sigma(a) = \int_a^b f(x)\, dx \qquad (2.2.4)$$

2. To determine the form of this conservation law *at the point $\bar{x}$*, we take the limit of each side of (2.2.4) as a approaches $\bar{x}$ from the left (written $a \longrightarrow \bar{x}^-$) and as b approaches $\bar{x}$ from the right ($b \longrightarrow \bar{x}^+$). Since the source terms f are bounded, the limit of the integral vanishes and we have

$$\lim_{b \to \bar{x}^+} \sigma(b) - \lim_{a \to \bar{x}^-} \sigma(a) = \lim_{\substack{a \to \bar{x}^- \\ b \to \bar{x}^+}} \int_a^b f(x)\, dx = 0$$

or

$$[\![\sigma(\bar{x})]\!] = 0 \qquad (2.2.5)$$

where

$$[\![\sigma(\bar{x})]\!] = \lim_{b \to \bar{x}^+} \sigma(b) - \lim_{a \to \bar{x}^-} \sigma(a)$$

is the *jump* in σ at $\bar{x}$. Thus, (2.2.5) states that, having no jumps, the flux is continuous at all points inside the smooth subdomains Ω_i, $i = 1, 2, 3, 4$.

3. Since the integrand $f(x)$ in (2.2.4) is continuous, we know from the mean-value theorem of integral calculus that $\int_a^b f(x)\, dx = (b - a)f(\zeta)$, ζ being some point such that $a < \zeta < b$ and $f(\zeta)$ being the average of f over this interval. Thus,

$$\sigma(b) - \sigma(a) = (b - a)f(\zeta)$$

Dividing by $b - a$ and taking the limit as before, we obtain

$$\lim_{\substack{a \to \bar{x}^- \\ b \to \bar{x}^+}} \frac{\sigma(b) - \sigma(a)}{b - a} = \lim_{\substack{a \to \bar{x}^- \\ b \to \bar{x}^+}} f(\zeta), \qquad a < \zeta < b$$

Because f is continuous, the right-hand side has a limit, $f(\bar{x})$, and so, therefore, does the left-hand side, namely $d\sigma(\bar{x})/dx$. Thus, we have, for all points x in the interior of the regions Ω_i,

$$\frac{d\sigma(x)}{dx} = f(x) \tag{2.2.6}$$

Substitution of the constitutive equation (2.2.3) into (2.2.6) produces a linear second-order differential equation for our boundary-value problem of the form

$$-\frac{d}{dx}\left[k(x) \frac{du(x)}{dx} \right] = f(x) \tag{2.2.7}$$

Of course, at points where the material modulus k is smooth, (2.2.7) can be expanded to read

$$-k(x)\frac{d^2u(x)}{dx^2} - \frac{dk(x)}{dx}\frac{du(x)}{dx} = f(x) \tag{2.2.8}$$

which, we recognize, is of the form (2.2.1) (with $a_0 = -k$, $a_1 = -dk/dx$, $a_2 = 0$).

4. We consider next the points at which some of the data are discontinuous but finite, such as the point $x = x_3$. As in the preceding discussion, the conservation principle requires that the jump in the flux vanishes; that is,

$$[\![\sigma(x)]\!] = 0, \qquad x = x_3$$

Because of the discontinuity in the integrand, we cannot use the mean-value theorem here; ku' is continuous at x_3 but $(ku')'$ is not defined. Hence, *we do not have a differential equation at the point* $x = x_3$.

At the point $x = x_1$ where f is continuous but k discontinuous, the conservation principle again gives a jump condition similar to (2.2.5); that is, $[\![\sigma(x)]\!] = 0$ for $x = x_1$. At this point we easily verify that the differential equation (2.2.6) must be satisfied, but since k is not differentiable at x_1, we cannot expand (2.2.7) to obtain the form given in (2.2.8).

5. Next we consider the point $x = x_2$ at which the concentrated source $f = \hat{f}\delta(x - x_2)$ is applied. Writing, as before, the balance of the flux equation for the region containing $x = x_2$, we have

$$\sigma(b) - \sigma(a) = \int_a^b f(x)\, dx = \int_a^b \bar{f}(x)\, dx + \int_a^b \hat{f}\delta(x - x_2)\, dx \quad (2.2.9)$$

where $\bar{f}$ is the smooth part of f and in the last integral we have employed the *symbolism* for Dirac deltas mentioned in Chapter 1 (recall equation (1.2.2)). Taking the limit as a and b approach x_2, we find that the integral of $\bar{f}$ goes to zero while the contribution due to the concentrated source has the constant value of $\hat{f}$, independent of a and b. Thus, we have the *non-homogeneous* jump condition

$$[\![\sigma(x)]\!] = \hat{f}, \qquad x = x_2$$

Since the value of the last integral is independent of a and b, we cannot obtain a differential equation for $x = x_2$.

6. Finally, we consider the boundaries of the body, i.e., $x = x_0$ and $x = x_4$. In physical situations, there is always the issue of definition of the physical system under investigation; all finite physical systems must interact in some way with their surrounding environment, and the effects of this environment are generally incorporated into a model of the system by prescribing the value of the flux or the state variable at the interface of the body and its surroundings, i.e., at the boundaries. Thus, in general, an approximate modeling of the medium adjacent to the boundary is used to provide the boundary condition.

    Figure 2.2c shows a small segment of the body containing the left end-point. The boundary flux at this point is denoted σ_0. Flux on this segment is conserved if

$$\sigma(a) - \sigma_0 = \int_0^a f(x)\, dx$$

Taking the limit as $a \to 0$ gives the condition $\sigma(0) = \sigma_0$. When we define the flux at a boundary point, we must recognize that a direction must be assigned to the rates-of-change of u at these points to reflect that the flux is entering or leaving the body. Specifically, if the flux is specified as $\sigma = \sigma_0$ and $\sigma = \sigma_l$ at the endpoints $x = 0$ and $x = l$, respectively, we take

$$\left.\begin{aligned} -k(0)\left(-\frac{du(0)}{dx}\right) &= \sigma_0 \\ -k(l)\frac{du(l)}{dx} &= \sigma_l \end{aligned}\right\} \quad (2.2.10)$$

wherein $du(0)/dx$ and $du(l)/dx$ are obviously one-sided derivatives. In many applications the flux at the boundary is known, so that (2.2.10) become boundary conditions. For second-order equations such as (2.2.7), boundary conditions such as (2.2.10), which involve the first derivatives of the unknown u, are called *natural boundary conditions*.

In another typical situation, the boundary flux is assumed to be proportional to the difference between the value of the state variable at the boundary and its value some distance into the adjacent medium. This amounts to assuming a simplified form of the constitutive equation for the adjacent medium. For example, at $x = 0$ this condition assumes the form

$$\sigma_0 = p_0[u(0) - u_0] \qquad (2.2.11)$$

In (2.2.11), p_0 is a known constant that depends upon the material modulus of the adjacent medium and u_0 is the known value of the state variable in the medium. Substitution of (2.2.11) into (2.2.10) gives the following form of boundary condition for this situation:

$$k(0) \frac{du(0)}{dx} = p_0[u(0) - u_0] \qquad (2.2.12)$$

Since $u'(0)$ appears here, (2.2.12) is again a form of a natural boundary condition for equation (2.2.7).

The choice of boundary conditions in a particular boundary-value problem is, of course, governed by the physics of the phenomena under study. In the physical problems considered here, this choice is made from among the following:

1. Essential boundary conditions, in which the value of state variable u at the boundary is specified.

2. Natural boundary conditions, in which either
    (a) the value of flux at the boundary is specified [(2.2.10)], or
    (b) a linear combination of the flux and the state variable at the boundary is specified [(2.2.12)].

We note that in case the physical problem requires flux boundary conditions at both ends, the prescribed values of flux must satisfy a global condition of conservation given by

$$\sigma_l + \sigma_0 = \int_0^l f(x)\, dx \qquad (2.2.13)$$

State-Dependent Sources and Transport Terms: For completeness, we note that there are two other terms which may appear in the governing differential equation. First is a term that arises when there is a source in the physical system which is proportional to the state variable u. Letting the constant of proportionality be the given function $-b(x)$, the differential equation becomes

$$-\frac{d}{dx}\left[k(x)\frac{du(x)}{dx}\right] + b(x)u(x) = f(x) \qquad (2.2.14)$$

Second, there is an important class of boundary-value problems which give rise to the remaining term in the general form of the differential equation: namely, steady-state transport processes. In these problems, the domain is viewed as a fixed region of space with material flowing through it. The flux quantity to be conserved is transmitted through the material, as in the discussion above, but is also carried, or convected, through the region of interest by the moving material. The extra term in the conservation equation arises from the calculation of the substantial, or total, time derivative of the state variable. The state variable u is then a function of position and time (i.e., $u = u(x, t)$), and the position x is a function of time (i.e., $x = x(t)$). Thus, the rate of change of u with respect to time is

$$\frac{du}{dt} = \frac{\partial u}{\partial t} + \frac{\partial u}{\partial x}\frac{\partial x}{\partial t}$$

When the process is steady in time, $\partial u/\partial t = 0$ and the velocity $\partial x/\partial t$ is constant (with respect to time) but may be different from zero. Let the velocity be the given function $c = c(x)$. Then the second term of the expression for du/dt gives a nonzero contribution to the conservation law of the form $c(du/dx)$, so that the governing differential equation is

$$-\frac{d}{dx}\left[k(x)\frac{du(x)}{dx}\right] + c(x)\frac{du(x)}{dx} + b(x)u(x) = f(x) \qquad (2.2.15)$$

which must be satisfied at all points where the data are smooth. Of course, we could consider cases in which discontinuities in c and b occur as well as in k and f. The jump conditions at interfaces then assume a more complicated form. Here we shall avoid such additional complexities for the sake of clarity by assuming that c and b are continuous throughout the domain Ω.

A General Two-Point Physical Problem: In summary, the application of the conservation principle, together with the constitutive law and prescribed conditions on the state variable u, leads to the following boundary-value problem: given the data k, c, b, f in the body and data such as

u_0, u_l, σ_0, σ_l, p_0, p_l at the boundaries, find the continuous function $u = u(x)$ that satisfies:

1. The governing differential equation at all points interior to the smooth subdomains,

$$-\frac{d}{dx}\left[k(x)\frac{du(x)}{dx}\right] + c(x)\frac{du(x)}{dx} + b(x)u(x) = f(x),$$
$$x \in \Omega_i, \quad i = 1, 2, 3, 4 \tag{2.2.16a}$$

2. The jump conditions at all points of discontinuous data,

$$\left[\!\left[k(x)\frac{du(x)}{dx}\right]\!\right] = 0, \qquad x = x_1$$
$$-\left[\!\left[k(x)\frac{du(x)}{dx}\right]\!\right] = \hat{f}, \qquad x = x_2 \qquad \Bigg\} \tag{2.2.16b}$$
$$\left[\!\left[k(x)\frac{du(x)}{dx}\right]\!\right] = 0, \qquad x = x_3$$

3. One of the following types of boundary conditions at each end of the interval:

$$u(x) = u_0 \text{ or } u_l \quad \text{at } x = 0 \text{ or } x = l$$
$$k(x)\frac{du(x)}{dx} = \sigma_0 \qquad \text{at } x = 0,$$
$$-k(x)\frac{du(x)}{dx} = \sigma_l \qquad \text{at } x = l$$
$$k(x)\frac{du(x)}{dx} - p_0 u(x) = -p_0 u_0 \qquad \text{at } x = 0 \qquad \text{or}$$
$$-k(x)\frac{du(x)}{dx} - p_l u(x) = -p_l u_l \qquad \text{at } x = l \tag{2.2.16c}$$

EXERCISES

2.2.1   Identify the coefficients $a_i(x)$ in (2.2.1) in terms of the physical data k, c, and b in (2.2.16a) and the boundary data α, β, and γ of (2.2.2) in terms of the physical boundary data in (2.2.16c).

2.2.2   Take $b = c = 0$ in (2.2.16a) and suppose that the flux σ is prescribed at points $x = 0$ and $x = l$. Prove that, in this case, condition (2.2.13) must hold if the conservation law is to hold for the entire body.

2.2.3 Suppose that $u(0) = u(l) = 0$, $b = c = 0$, $f(x) = f_0 = $ constant, and

$$k(x) = \begin{cases} k_1, & 0 \le x < \dfrac{l}{2} \\[2mm] k_2, & \dfrac{l}{2} < x \le l \end{cases}$$

Calculate the exact solution $u = u(x)$ of the boundary-value problem (2.2.16) and verify by direct calculation that $[\![k(x)u'(x)]\!] = 0$ at $x = l/2$.

2.3 VARIATIONAL FORMULATION OF TWO-POINT BOUNDARY-VALUE PROBLEMS

Now it should be clear that the "classical" statement of a boundary-value problem, such as that represented by (2.2.1) and (2.2.2), suggests conditions on the regularity of the solution which are too strong to make sense for many reasonable choices of the data. What we want as a basis for finite element approximations is a much richer and weaker variational statement of the problem. Moreover, to provide some symmetry in the formulation,* we generally prefer to choose a formulation in which the trial functions u and the test function v have the same degree of smoothness; that is, we would like to have a variational statement in which the highest order of derivatives of u that appears is the same as that of v. To demonstrate how this is done, consider a boundary-value problem of the type described in (2.2.16a)–(2.2.16c) defined by the system

$$\left. \begin{aligned} -\frac{d}{dx}\left[k(x)\frac{du(x)}{dx}\right] + c(x)\frac{du(x)}{dx} + b(x)u(x) = f(x) \\ \text{for points } x \in \Omega_i, \quad i = 1, 2, 3, 4 \\ \left[\!\!\left[k(x)\frac{du(x)}{dx} \right]\!\!\right] = 0 \quad \text{at } x = x_1 \\ -\left[\!\!\left[k(x)\frac{du(x)}{dx} \right]\!\!\right] = \hat{f} \quad \text{at } x = x_2 \\ \left[\!\!\left[k(x)\frac{du(x)}{dx} \right]\!\!\right] = 0 \quad \text{at } x = x_3 \\ \alpha_0\frac{du(0)}{dx} + \beta_0 u(0) = \gamma_0, \quad \alpha_l\frac{du(l)}{dx} + \beta_l u(l) = \gamma_l \end{aligned} \right\} \quad (2.3.1)$$

wherein the domain Ω is divided into the four smooth subdomains Ω_i indi-

---

* Here we refer to symmetry in the smoothness requirements on u and v and not to symmetry in the final stiffness matrix, which often is not possible.

cated in Fig. 2.1 by the interface points $x_0 = 0$, x_1, x_2, x_3, and $x_4 = l$. For example, for the third type of condition in (2.2.16c), possible choices of $\alpha_0, \beta_0, \ldots, \gamma_l$ are

$$\alpha_0 = k(0), \quad \beta_0 = -p_0, \quad \gamma_0 = -p_0 u_0, \quad \alpha_l = -k(l), \quad \beta_l = -p_l, \quad \gamma_l = -p_l u_l$$

Now we can arrive quite easily at a variational statement of this problem. First note that the solution u is quite smooth inside each of the subdomains Ω_i. Indeed, by virtue of the fact that the differential equation in (2.3.1) must hold in these subdomains, u is at least twice differentiable there. Following the procedure described in Chapter 1, we construct the residual error function r, where (again denoting derivatives by primes)

$$r(x) = -[k(x)u'(x)]' + c(x)u'(x) + b(x)u(x) - f(x),$$
$$x \in \Omega_i, \quad i = 1, 2, 3, 4$$

multiply r by a sufficiently smooth test function v defined over the entire interval, $0 < x \le l$, and integrate the first term in the product rv by parts over each subdomain. The result over subdomain Ω_i is of the form

$$\int_{\Omega_i} rv \, dx = -ku'v \Big|_{x_{i-1}}^{x_i} + \int_{\Omega_i} (ku'v' + cu'v + buv) \, dx - \int_{\Omega_i} fv \, dx, \tag{2.3.2}$$
$$i = 1, 2, 3, 4$$

Now, since u is the solution of our problem,

$$\int_{\Omega_i} rv \, dx = 0 \quad \text{and, therefore,} \quad \sum_{i=1}^{4} \int_{\Omega_i} rv \, dx = 0 \tag{2.3.3}$$

Thus, substituting (2.3.2) into (2.3.3) yields

$$\int_0^l (ku'v' + cu'v + buv) \, dx + k(0)u'(0)v(0) + [\![k(x_1)u'(x_1)]\!]v(x_1)$$
$$+ [\![k(x_2)u'(x_2)]\!]v(x_2) + [\![k(x_3)u'(x_3)]\!]v(x_3) - k(l)u'(l)v(l) \tag{2.3.4}$$
$$= \int_0^l \bar{f}v \, dx$$

for all smooth test functions v. As in the preceding section,

$$[\![k(x_i)u'(x_i)]\!] = \lim_{x \to x_i^+} k(x)u'(x) - \lim_{x \to x_i^-} k(x)u'(x)$$

and the function $\bar{f}$ appearing on the right-hand side of (2.3.4) is understood to be the "smooth part" (i.e., the integrable part) of the source function f.

In view of (2.3.1),

$$[\![k(x_1)u'(x_1)]\!] = 0, \qquad -[\![k(x_2)u'(x_2)]\!] = \hat{f}$$

$$[\![k(x_3)u'(x_3)]\!] = 0, \qquad u'(0) = \frac{\gamma_0 - u(0)\beta_0}{\alpha_0}$$

$$u'(l) = \frac{\gamma_l - u(l)\beta_l}{\alpha_l}$$

Thus, (2.3.4) reduces to

$$\int_0^l (ku'v' + cu'v + buv)\, dx$$

$$= \int_0^l \bar{f}v\, dx + \hat{f}v(x_2) - \frac{k(0)}{\alpha_0}[\gamma_0 - u(0)\beta_0]v(0) \qquad (2.3.5)$$

$$+ \frac{k(l)}{\alpha_l}[\gamma_l - u(l)\beta_l]v(l)$$

for all admissible test functions v.

A variational statement of the two-point boundary-value problem (2.3.1) now takes on the following form: *find a function u such that* (2.3.5) *holds for all test functions v in a suitable class of admissible functions.*

This is a rather remarkable result. We have managed to transform the entire system of differential equations, jump conditions, and boundary conditions in (2.3.1) into a single equation in which all of the features of the solution and the discontinuous data are intrinsically present. The variational problem (2.3.5) characterizes the solution as a function defined over the entire interval, $0 \le x \le l$, rather than piecewise as in (2.3.1). Nevertheless, it is clear that any solution of (2.3.1) is automatically a solution of (2.3.5). As in Chapter 1, we henceforth view the variational statement (2.3.5) as a *given* variational problem. It will always include the classical problem as a special case whenever the solution is sufficiently smooth.

Problem (2.3.5) is still incompletely defined. Our study of the model problem in Chapter 1 established that the specification of the appropriate space of admissible functions lies at the heart of our analysis. In addition, the character of the boundary terms in (2.3.5) deserves further comment. We now list several fundamentally important observations which lead to a more concrete definition of the variational statement of our problem.

1.   By integrating $(ku')'v$ once by parts, we have produced an integral involving products of the first derivatives of trial functions u and test functions v. Thus, if we wish to identify a class of admissible functions on which smoothness assumptions are barely strong enough to make this integral well defined, it is sufficient to take u and v to be

members of a class of functions, denoted H^1, whose derivatives of order 1 and less are square-integrable over Ω. In other words, a test function v will belong to H^1 if

$$\int_0^l [(v')^2 + v^2]\, dx < +\infty \qquad (2.3.6)$$

With these conventions, it is clear that the variational statement of problem (2.3.1) is as follows:

$$\left.\begin{array}{l} \text{Find a function } u \in H^1 \text{ such that (2.3.5) holds} \\ \text{for all test functions } v \in H^1. \end{array}\right\} \qquad (2.3.7)$$

We also encounter frequently the subclass of functions in H^1 that vanish at $x = 0$ and $x = l$. We denote this class by H_0^1; that is, $v = v(x)$ is a member of H_0^1 if

(a) v satisfies (2.3.6).

(b) $v(0) = 0$; $v(l) = 0$.

In particular, consider the case in which we have *essential* boundary conditions of the form

$$u(0) = \frac{\gamma_0}{\beta_0} \qquad \text{and} \qquad u(l) = \frac{\gamma_l}{\beta_l} \qquad (2.3.8)$$

instead of the natural boundary conditions in (2.3.1). Then the variational boundary-value problem becomes:

$$\left.\begin{array}{c} \text{Find a function } u \text{ in } H^1 \text{ satisfying (2.3.8) such that} \\[2mm] \int_0^l (ku'v' + cu'v + buv)\, dx = \int_0^l \bar{f}v\, dx + \hat{f}v(x_2) \\[2mm] \text{for all } v \in H_0^1. \end{array}\right\} \qquad (2.3.9)$$

2.   Boundary conditions for a problem such as (2.3.1) cannot be arbitrarily constructed; they must be, in some sense, compatible with the governing differential equation of the problem. For example, the specification of an arbitrary set of boundary conditions to define a specific solution of a differential equation may lead to an "ill posed" problem in which the solution does not exist at all or exists but is not uniquely defined or "well-behaved" (see Exercise 2.3.2). Another useful feature of the variational formulation is that whenever the boundary conditions can be incorporated naturally into an integration-by-parts formula, as was the case in our derivation of (2.3.5), they are automatically compatible with the differential equation.

Thus, there are intrinsic features of the variational statement of a boundary-value problem that serve to characterize well-posed problems.

3. Recall that in our discussion of physical origins of two-point problems in the preceding section, boundary conditions fell into two categories: *essential boundary conditions*, in which the value of the solution u is specified, and *natural boundary conditions*, in which u' or a combination of u' and u is specified. Exactly the same classification arises naturally in variational formulations such as (2.3.5). Suppose that u and v are in the class H^1. Then derivatives of u and v of order 2 and higher may not exist. If v is barely smooth enough to be in H^1, *it is impossible to impose conditions on derivatives of v of order 1 or higher* (see Exercise 2.3.3). From this observation, it follows that boundary conditions enter variational boundary-value problems of the type in (2.3.5) in two distinct ways: the essential boundary conditions, which involve the specification of values of the solution, enter the problem in the definition of the space of admissible functions, whereas the natural boundary conditions, which involve the specification of derivatives of the solution, dictate the actual form of the variational equation. In particular, we see from the form of problem (2.3.5) that natural boundary conditions appear on the right-hand side of the variational equality (2.3.5).

To fix ideas, consider as examples the problems

(i) $-u''(x) + u(x) = f(x), \qquad 0 < x < l$

$$u(0) = 0, \qquad u(l) = 0$$

(ii) $-u''(x) + u(x) = f(x), \qquad 0 < x < l$

$$u'(0) = \gamma_0, \qquad u'(l) = \gamma_l$$

According to (2.3.5) and (2.3.9), the variational statements of these problems are:

(V-i)   Find u in H_0^1 such that

$$\int_0^1 (u'v' + uv)\, dx = \int_0^1 fv\, dx \qquad \text{for all } v \in H_0^1$$

wherein H_0^1 is the class of functions v satisfying (2.3.6) and vanishing at the boundaries: $v(0) = 0 = v(l)$.

(V-ii)   Find $u \in H^1$ such that

$$\int_0^1 (u'v' + uv)\, dx = \int_0^1 fv\, dx - \gamma_0 v(0) + \gamma_l v(l)$$

$$\text{for all } v \in H^1$$

Clearly, the essential boundary conditions in (V-i) enter the variational problem through the definition of the class H_0^1 of admissible test functions, whereas the natural boundary conditions in (V-ii) enter as data on the right-hand side of the equation.

4. As a final remark, we note that, in analogy with the function U in equation (1.7.1), we may also define as the *energy norm* for problem (2.3.5) when $b(x) > 0$,

$$\| v \|_E = \left[\int_0^l (kv'^2 + bv^2) \, dx \right]^{1/2} \qquad (2.3.10)$$

Frequently, we employ the equivalent H^1-norm

$$\| v \|_1 = \left[\int_0^l (v'^2 + v^2) \, dx \right]^{1/2} \qquad (2.3.11)$$

This norm provides a natural measure of error in approximations to problems like (2.3.5).

EXERCISES

2.3.1 Construct variational statements of the following boundary-value problems, indicating in each case the appropriate space of admissible test functions.

(a) $-[k(x)u'(x)]' + u'(x) + u(x) = 0$ for $0 < x < 1$, $1 < x < 2$, $2 < x < 3$, $3 < x < 4$

$$[\![k(1)u'(1)]\!] = [\![k(2)u'(2)]\!] = 0, \qquad [\![k(3)u'(3)]\!] = 10$$

$$k(x) = \begin{cases} 1 & 0 \le x < 1 \\ 2 & 1 \le x < 2 \\ 1 & 2 \le x \le 4 \end{cases}$$

$$u(0) = 0, \qquad u'(4) = 3$$

(b) $-u''(x) + u(x) = \delta(x - 1)$, $0 < x < 2$

$$u'(0) = 2, \qquad u'(2) + u(2) = 3$$

(c) $-u''(x) + u'(x) = 0$, $0 < x < 1$ and $1 < x < 2$

$$[\![u'(1)]\!] = 1, \qquad u(0) = u(2) = 0$$

2.3.2 Consider the boundary-value problem

$$-u''(x) + u(x) = x, \qquad 0 < x < 1$$

$$a_{11}u'(0) + a_{12}u(0) + a_{13}u'(1) + a_{14}u(1) = b_1$$

$$a_{21}u'(0) + a_{22}u(0) + a_{23}u'(1) + a_{24}u(1) = b_2$$

where $a_{11}, a_{12}, \ldots, a_{23}, a_{24}, b_1$, and b_2 are constants. Determine conditions on the constants appearing in the boundary conditions in order that these conditions be compatible with the differential equation in the sense discussed in the text. In particular, show that it is sufficient to require that the determinant

$$\begin{vmatrix} a_{11} & a_{13} \\ a_{21} & a_{23} \end{vmatrix} \neq 0$$

2.3.3 Consider the function u defined on $0 \leq x \leq 1$ which is barely smooth enough to be in H^1:

$$u(x) = \begin{cases} x & 0 < x \leq \frac{1}{2} \\ \frac{1}{2} - x & \frac{1}{2} \leq x \leq 1 \end{cases}$$

with $u(0) = 0$, $u(1) = 0$. Observe that u has a "corner" at $x = \frac{1}{2}$; u' is, thus, not uniquely defined at $x = \frac{1}{2}$ even though $(u')^2$ is integrable. On the basis of this result, discuss why it is impossible to specify the value of the derivative u' of a function $u \in H_0^1$ at an arbitrary point x_0, $0 < x_0 < 1$.

2.3.4 The following boundary-value problems are ill-posed or do not satisfy the assumptions described in the text. What is wrong with them?

(a) $\left(\dfrac{1}{x-1}\right)u''(x) + u(x) = \sin x$, $0 < x < 2$

$$u(0) = 0 = u(2)$$

(b) $(1 - x^2)u''(x) + u'(x) = 3$, $0 < x < 2$

$$u'(0) = 0, \qquad u'(2) = 1$$

(c) $u''(x) + xu(x) = 3$, $0 < x < 1$

$$u''(0) = 0, \qquad u(1) = 0$$

(d) $-u''(x) + u(x) = x$, $0 < x < 1$

$$\tfrac{1}{2}u'(0) + u'(1) = u(1), \qquad u'(0) + 2u'(1) = u(0)$$

2.4 GALERKIN APPROXIMATIONS

The Galerkin approximation of second-order boundary-value problems follows exactly the same lines as those discussed for the model problem in Chapter 1. We identify a finite set of basis functions $\{\phi_1, \phi_2, \ldots, \phi_N\}$ in H^1 that define a finite-dimensional subspace of test functions H^h in H^1. We then seek a function $u_h \in H^h$ of the form

$$u_h(x) = \sum_{j=1}^{N} \alpha_j \phi_j(x) \tag{2.4.1}$$

which satisfies the variational problem on H^h. For problem (2.3.5), this procedure leads to the discrete problem

$$
\int_0^l (ku_h'v_h' + cu_h'v_h + bu_hv_h)\, dx = \int_0^l \bar{f}v_h\, dx + \hat{f}v_h(x_2)
$$

$$
- \frac{k(0)}{\alpha_0}[\gamma_0 - u_h(0)\beta_0]v_h(0) + \frac{k(l)}{\alpha_l}[\gamma_l - u_h(l)\beta_l]v_h(l) \tag{2.4.2}
$$

$$
\text{for all } v_h \in H^h
$$

or, equivalently,

$$
\sum_{j=1}^N K_{ij}\alpha_j = F_i, \qquad i = 1, 2, \ldots, N \tag{2.4.3}
$$

where the stiffness matrix K_{ij} is now of the form

$$
K_{ij} = \int_0^l (k\phi_i'\phi_j' + c\phi_i'\phi_j + b\phi_i\phi_j)\, dx
$$
$$
- \frac{k(0)\beta_0}{\alpha_0}\phi_i(0)\phi_j(0) + \frac{k(l)\beta_l}{\alpha_l}\phi_i(l)\phi_j(l) \tag{2.4.4}
$$

and the components of the load vector are

$$
F_i = \int_0^l \bar{f}\phi_i\, dx + \hat{f}\phi_i(x_2) - \frac{k(0)}{\alpha_0}\gamma_0\phi_i(0) + \frac{k(l)}{\alpha_l}\gamma_l\phi_i(l) \tag{2.4.5}
$$

with $1 \leq i, j \leq N$.

Upon solving (2.4.3) for the coefficients α_j, our Galerkin approximation of the problem is obtained immediately from (2.4.1). As before, this method of approximation becomes very powerful whenever we have a systematic technique for constructing the basis functions ϕ_i. Such a systematic method is provided, of course, by finite element techniques, and we shall take this up in some detail in Sections 2.6 through 2.9.

We note that the stiffness matrix K_{ij} is not symmetric whenever the coefficient c in (2.4.4) is not identically zero.

EXERCISES

2.4.1 Construct a Galerkin approximation of (2.3.5) for the case in which $f(x) = 2$, $k(x) = -1$, $c(x) = 0$, $b(x) = 1$, $\alpha_0 = 1$, $\alpha_l = 1$, $\beta_0 = 0$, $\beta_l = 1$, $\gamma_0 = 1$, and $\gamma_l = 0$. Use a two-degree-of-freedom two-term approximation with $\phi_1(x) = 1 + \frac{1}{2}x^2$ and $\phi_2(x) = \frac{1}{3}x^3$. Are ϕ_1 and ϕ_2 admissible functions? Compare $u_h(\frac{1}{2})$ with the exact solution u at $x = 1/2$. (Take $l = 1$.)

2.4.2　Formulate the equations for the Galerkin approximation of the variational problem (2.3.9) with essential boundary conditions.

2.5 MINIMIZATION OF ENERGY FUNCTIONALS*

Our reference to certain weak forms of boundary-value problems as "variational" statements arises from the fact that, whenever the operators involved possess a certain symmetry, to be identified below, a weak form of the problem can be obtained which is precisely that arising in standard problems in the calculus of variations. In such cases, the variational boundary-value problem represents a characterization of the function u which minimizes (or maximizes) the energy of the problem. Since one may find this interpretation useful from time to time, we summarize here some of these variational concepts.

Let us consider once again a class H_0^1 of functions v defined on the interval $0 < x < l$ and vanishing at the endpoints. Suppose that J denotes a real-valued function defined on H_0^1 given by

$$J(v) = \tfrac{1}{2} \int_0^l (kv'^2 + bv^2 - 2fv)\,dx \tag{2.5.1}$$

where k, b, and f are given functions of x, with k and b satisfying $0 < k_0 \leq k(x) < \infty$ and $b(x) \geq 0$ for all x, k_0 being a positive constant. We may generally regard J as the *energy* of a certain physical system. Note that J is a "function of functions" (i.e., the domain of J is the class H_0^1 of admissible functions) and that the values of J are real numbers. Any function with these properties is called a *functional*.

Now a classical minimization problem in the calculus of variations is to seek a particular function $u \in H_0^1$ at which J assumes its smallest value over the whole class H_0^1. In other words, "u is the minimizer of J over H_0^1" means that

$$J(u) \leq J(v) \qquad \text{for all } v \in H_0^1 \tag{2.5.2}$$

The minimization problem now reduces to one of characterizing the minimizing function u. Toward this end, we consider an arbitrary function $\eta \in H_0^1$ of the form $\eta = u + \epsilon v$, v arbitrary, where ϵ is a positive number. Then η can be made as close to u as possible by choosing ϵ small enough. The "perturbation" ϵv in u is called a *variation* in u and is often written δu.

---

* The ideas discussed here are not used in subsequent sections and can be skipped without serious consequence.

The value of the energy at η is

$$J(\eta) = J(u + \epsilon v)$$
$$= J(u) + \epsilon \delta J(u; v) + \epsilon^2 \delta^2 J(v) \tag{2.5.3}$$

where, by a direct calculation using (2.5.1),

$$\delta J(u; v) = \int_0^l (ku'v' + buv - fv) \, dx \tag{2.5.4}$$

and

$$\delta^2 J(v) = \tfrac{1}{2} \int_0^l (kv'^2 + bv^2) \, dx \tag{2.5.5}$$

The quantity $\delta J(u; v)$ is called the *first variation in J at u* and $\delta^2 J(v)$ is called the *second variation in J at u*. It is clear that $\delta J(u; v)$ can also be calculated using the formula

$$\delta J(u; v) = \lim_{\epsilon \to 0} \frac{1}{\epsilon} [J(u + \epsilon v) - J(u)]$$
$$= \frac{\partial}{\partial \epsilon} J(u + \epsilon v)|_{\epsilon=0} \tag{2.5.6}$$

Now since $\delta^2 J(v)$ is, because of our assumptions on k and b, always greater than or equal to zero, and since u is the minimizer of J,

$$J(u) \le J(u + \epsilon v) = J(u) + \epsilon \delta J(u; v) + \epsilon^2 \delta^2 J(v)$$

Thus,

$$\delta J(u; v) + \epsilon \delta^2 J(v) \ge 0$$

or, taking the limit as ϵ goes to zero,

$$\delta J(u; v) \ge 0 \qquad \text{for all } v \in H_0^1$$

But this inequality must also hold if v is replaced by $-v$, so that if u is the actual minimizer of J, we must have

$$\delta J(u; v) = 0 \qquad \text{for all } v \in H_0^1 \tag{2.5.7}$$

In other words, the minimizer of J is characterized as the solution of the variational boundary-value problem

$$\int_0^l (ku'v' + buv) \, dx = \int_0^l fv \, dx \qquad \text{for all } v \in H_0^1 \tag{2.5.8}$$

We recognize (2.5.8) as a variational statement of the classical boundary-value problem

$$-[k(x)u'(x)]' + b(x)u(x) = f(x), \qquad 0 < x < l$$
$$u(0) = 0, \qquad u(l) = 0 \tag{2.5.9}$$

These concepts drawn from variational calculus are the basis for our use of the term "variational boundary-value problem" when we refer to the weak statement (2.5.8) of (2.5.9). We continue to refer to problems of the form (2.5.8) as variational problems even in those cases in which they cannot be derived from a problem of minimizing some energy functional.

It is also clear that the problem of approximating the solution u of (2.5.8) can be approached as one of seeking a minimizer of J in some finite-dimensional subspace H^h of H_0^1. Then, if $u_h(x) = \sum_{i=1}^{N} u_i \phi_i(x)$,

$$
\begin{aligned}
J(u_h) &= \tfrac{1}{2} \int_0^l [k(u_h')^2 + bu_h^2 - 2fu_h] \, dx \\
&= \tfrac{1}{2} \sum_{i=1}^{N} \sum_{j=1}^{N} u_j \left[\int_0^l (k\phi_i'\phi_j' + b\phi_i\phi_j - 2f\phi_i) \, dx \right] u_i \\
&= \tfrac{1}{2} \sum_{i=1}^{N} \sum_{j=1}^{N} u_i K_{ij} u_j - \sum_{i=1}^{N} F_i u_i
\end{aligned} \tag{2.5.10}
$$

where K_{ij} and F_i are the components of the stiffness matrix and the load vector, respectively:

$$K_{ij} = \int_0^l (k\phi_i'\phi_j' + b\phi_i\phi_j) \, dx; \qquad F_i = \int_0^l f\phi_i \, dx \tag{2.5.11}$$

$J(u_h)$ is minimized by choosing the coefficients u_i, so that

$$\frac{\partial J(u_h)}{\partial u_i} = 0; \qquad i = 1, 2, \ldots, N \tag{2.5.12}$$

Thus, once again, we arrive at the system of equations

$$\sum_{j=1}^{N} K_{ij} u_j = F_i; \qquad i = 1, 2, \ldots, N \tag{2.5.13}$$

The approximation scheme outlined above is called the *Ritz method*. We see that the Ritz method can be used as a basis for constructing finite element approximations of variational boundary-value problems whenever the problem is equivalent to finding a function u which makes the first variation of an energy functional J vanish for all admissible variations v.

We conclude with a final observation of some importance. Note that the governing differential equation in linear second-order problems can be written compactly in the operator form

$$Au = f \qquad (2.5.14)$$

where A is the differential operator for the problem. If u and v are arbitrary smooth functions vanishing at $x = 0$ and $x = l$, the operator A is said to be *formally self-adjoint* whenever

$$\int_0^l v\, Au\; dx = \int_0^l u\, Av\; dx \qquad (2.5.15)$$

It can be shown that an energy functional J of the type in (2.5.1) exists for a given boundary-value problem only when the operator A for the problem is self-adjoint. For self-adjoint problems and, therefore, for all problems derivable from an energy functional in the manner outlined above, the stiffness matrix (2.5.11) resulting from a Ritz approximation will always be symmetric. Clearly, when Ritz's method is applicable, it leads to the same system of equations as Galerkin's method. In the general case discussed in Section 2.4, the operator was not self-adjoint. For this reason, it is clear that Galerkin's method is applicable to a wider class of problems than is Ritz's method.

EXERCISES

2.5.1 Derive variational boundary-value problems that characterize minima of the following energy functionals:

(a)
$$J(v) = \tfrac{1}{2}\int_0^l [(v'')^2 - 2(v')^2 + v^2 - 2fv]\, dx$$
$$v(0) = v'(0) = 0, \qquad v(1) = v'(1) = 0$$

(b)
$$J(v) = \tfrac{1}{2}\int_0^l [(v')^2 + v^2 - 2fv]\, dx$$
$$v(0) = v(1) = 0$$

(c)
$$J(v) = \tfrac{1}{2}\int_0^l (kv'^2 + bv^2 - 2fv)\, dx$$
$$+ \tfrac{1}{2}\beta_0[v(0)]^2 - v(0)\gamma_0$$
$$+ \tfrac{1}{2}\beta_l[v(l)]^2 - v(l)\gamma_l$$

In all cases, describe the appropriate spaces of admissible functions.

2.5.2 Let J denote the functional

$$J(v) = \int_0^l F(v(x), v'(x))\, dx$$

where F is a continuously differentiable function of its arguments. Show that if the test functions v are sufficiently smooth, a minimizer u of J will satisfy the differential equation

$$\frac{\partial F(u, u')}{\partial u} - \frac{d}{dx}\left[\frac{\partial F(u, u')}{\partial u'}\right] = 0, \qquad 0 < x < l$$

Discuss possible natural boundary conditions for this problem.

2.5.3 Which of the following operators are formally self-adjoint?

(a) $Au(x) = -u''(x) + u(x), \qquad 0 < x < 1$

(b) $Au(x) = -u''(x) + u'(x) + u(x), \qquad 0 < x < 1$

(c) $Au(x) = (2 - x)u''(x) - u'(x) + u(x), \qquad 0 < x < 1$

2.5.4 Derive conditions on k, b, and c in (2.2.16a) which guarantee that A be formally self-adjoint.

2.6 FINITE ELEMENT INTERPOLATION

While we have presented the finite element method as a technique for systematically applying Galerkin's method to the approximate solution of boundary-value problems, a brief reflection reveals that the underlying ideas also provide a basis for methods of *interpolation*. Indeed, the finite element concepts can be used to construct curve-fitting schemes wherein any given function g can be approximated by a system of piecewise polynomials, the values* of which coincide with those of g at prescribed nodal points in the domain of g. When viewed in this way, a variety of choices of element shape functions come to mind which are merely bases of well-known methods of interpolating smooth functions.

Suppose that we are given a function g defined on an interval $0 \le x \le l$ and that g is smooth enough to be continuously differentiated k times and that its derivative of order $k + 1$ is bounded (finite) on this interval. We wish to construct a finite element approximation (an interpolant) g_h of g that coincides with g at the nodal points, and we wish to estimate the accuracy of such approximations. We begin, of course, by partitioning the interval into a

---

* We can, of course, also match values of derivatives of g of various orders at the nodes, and such higher-order interpolations are necessary for problems of order four and higher. We discuss such higher-order elements in Chapter 6. See also Exercise 2.6.6.

collection of finite elements. Then comes the problem of showing just how general shape functions can be constructed. We describe next a technique for generating polynomial shape functions of any degree k (i.e., each shape function ψ_i^e will contain monomials in x up to x^k, k being a positive integer). The technique leads to the *Lagrange families* of finite elements, the name "Lagrange" being borrowed from the notion of Lagrange interpolation, from which these element families are derived.

A Lagrange finite element employing polynomials of degree k is constructed as follows:

1. As usual, we consider a typical finite element Ω_e, isolated from the mesh, and we establish a local coordinate system ξ, with origin now at the center of the element, scaled so that $\xi = -1$ at the left endpoint and $\xi = 1$ at the right endpoint, as shown in Fig. 2.3a. This is achieved by the simple linear stretching transformation for the general element Ω_e,

    $$\xi = \frac{2x - (x_1^e + x_{k+1}^e)}{x_{k+1}^e - x_1^e} \tag{2.6.1}$$

    so that points x such that $x_1^k \leq x \leq x_{k+1}^k$ are transformed to points ξ such that $-1 \leq \xi \leq 1$. We perform our element calculations on this reference or "master" element $\hat{\Omega}$ and denote the shape functions on the master element by $\hat{\psi}_i(\xi)$.

2. For shape functions of degree k, we identify $k + 1$ nodes (including the endpoints) which divide the element into k equal segments. Let ξ_i, $i = 1, 2, \ldots, k + 1$, denote the ξ-coordinates of each node. For each node ξ_i, we form the product of k linear functions $(\xi - \xi_j)$, $j = 1, 2, \ldots, k + 1$, $j \neq i$. Note that this product is zero at all nodes except i. These functions are of the form

    node 1: $(\xi - \xi_2)(\xi - \xi_3) \ldots (\xi - \xi_{k+1})$

    node 2: $(\xi - \xi_1)(\xi - \xi_3)(\xi - \xi_4) \ldots (\xi - \xi_{k+1})$

    $\cdot$
    $\cdot$
    $\cdot$

    node i: $(\xi - \xi_1) \ldots (\xi - \xi_{i-1})(\xi - \xi_{i+1}) \ldots (\xi - \xi_{k+1})$

    $\cdot$
    $\cdot$
    $\cdot$

    node $k + 1$: $(\xi - \xi_1)(\xi - \xi_2) \ldots (\xi - \xi_k)$

3. For each node i, we evaluate the corresponding product in step 2 at $\xi = \xi_i$ and divide the product functions by this value. This normalizes

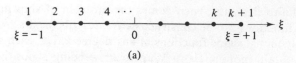

(a)

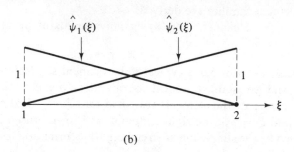

(b)

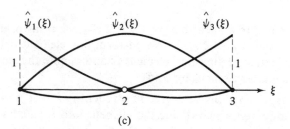

(c)

FIGURE 2.3 (a) A master element $\hat{\Omega}$ with k + 1 nodes; (b) linear shape functions corresponding to k = 1 and (c) an element with three nodes and piecewise quadratic shape functions (k = 2).

the polynomials so that $\hat{\psi}_i(\xi_i) = 1$ and produces the correct shape function $\hat{\psi}_i(\xi)$ corresponding to each node i. For example,

$$\hat{\psi}_1(\xi) = \frac{(\xi - \xi_2)(\xi - \xi_3)\ldots(\xi - \xi_{k+1})}{(\xi_1 - \xi_2)(\xi_1 - \xi_3)\ldots(\xi_1 - \xi_{k+1})}$$

$$\hat{\psi}_2(\xi) = \frac{(\xi - \xi_1)(\xi - \xi_3)\ldots(\xi - \xi_{k+1})}{(\xi_2 - \xi_1)(\xi_2 - \xi_3)\ldots(\xi_2 - \xi_{k+1})}$$

or, in general,

$$\hat{\psi}_i(\xi) = \frac{(\xi - \xi_1)(\xi - \xi_2)\ldots(\xi - \xi_{i-1})(\xi - \xi_{i+1})\ldots(\xi - \xi_{k+1})}{(\xi_i - \xi_1)(\xi_i - \xi_2)\ldots(\xi_i - \xi_{i-1})(\xi_i - \xi_{i+1})\ldots(\xi_i - \xi_{k+1})}$$

$$(2.6.2)$$

These functions have the property that

$$\hat{\psi}_i(\xi_j) = \begin{cases} 1 & \text{if } i = j \\ 0 & \text{if } i \neq j \end{cases} \qquad (2.6.3)$$

which implies that $\hat{\psi}_i(\xi)$ are linearly independent. These $k + 1$ functions define a basis for the set of all polynomials of degree k and we say that the basis $\hat{\psi}_i$ is complete.

This implies that any polynomial of degree k or less can be represented uniquely in terms of the Lagrange polynomial basis. This property carries over to the global basis functions ϕ_i: every polynomial of degree $\leq k$ can be expressed in a unique way as a linear combination of the basis functions ϕ_i generated using the Lagrange shape functions in (2.6.2).

Note that for $k = 1$ (linear shape functions), we have two nodes and the shape functions (see Fig. 2.3b) are

$$\left. \begin{array}{l} \hat{\psi}_1(\xi) = \dfrac{\xi - \xi_2}{\xi_1 - \xi_2} = \dfrac{1}{2}(1 - \xi) \\[2mm] \hat{\psi}_2(\xi) = \dfrac{\xi - \xi_1}{\xi_2 - \xi_1} = \dfrac{1}{2}(1 + \xi) \end{array} \right\} \qquad (2.6.4)$$

If we introduce the change of coordinates $\xi = 2\bar{x}/h - 1$, where $\bar{x} = x - x_i$, then $\psi_1^e(x) = 1 - \bar{x}/h$ and $\psi_2^e(x) = \bar{x}/h$, which we recognize as precisely the functions employed in Chapter 1 (recall Fig. 1.8). Upon connecting elements together to form the finite element mesh, the element functions match up to produce piecewise-linear basis functions ϕ_i of the form indicated in Fig. 2.4.

For $k = 2$ (quadratic shape functions), we have three nodes and the shape functions (see Fig. 2.3c)

$$\hat{\psi}_1(\xi) = \tfrac{1}{2}\xi(\xi - 1), \qquad \hat{\psi}_2(\xi) = 1 - \xi^2, \qquad \hat{\psi}_3(\xi) = \tfrac{1}{2}\xi(\xi + 1) \qquad (2.6.5)$$

The corresponding global basis functions ϕ_i are shown in Fig. 2.5.

An estimate of the error for piecewise-linear Lagrange interpolation can be obtained using Taylor series. Let $E = g - g_h$ be the interpolation error function and consider an arbitrary element Ω_e with points $x_i \leq x \leq x_{i+1}$ in the mesh. We assume that g has bounded second derivatives. Now, on Ω_e, $E = g - g_h$ can be expanded in a local Taylor series about any interior point $\bar{x}$:

$$E(x) = E(\bar{x}) + E'(\bar{x})(x - \bar{x}) + \tfrac{1}{2}E''(\zeta)(x - \bar{x})^2 \qquad (2.6.6)$$

where ζ is a point between $\bar{x}$ and x.

Since g_h is the interpolant of g, the error E is zero at the endpoints x_i, x_{i+1}.

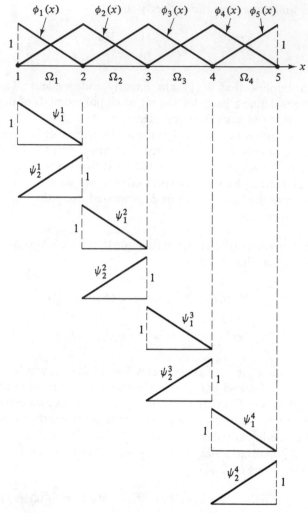

FIGURE 2.4 *Piecewise-linear basis functions ϕ_i for a 4-element mesh generated by linear shape functions, ψ_1^e, ψ_2^e defined over each element.*

We next select $\bar{x}$ to be that point at which $|E|$ is a maximum. At this point, $E'(\bar{x}) = 0$ so that (2.6.6) reduces to

$$E(x) = E(\bar{x}) + \tfrac{1}{2}E''(\zeta)(x - \bar{x})^2 \qquad (2.6.7)$$

for $x_i \leq x \leq x_{i+1}$. Next, we set $x = x_i$ or x_{i+1} whichever is closer to $\bar{x}$ (say x_i). Then

$$E(\bar{x}) = -\tfrac{1}{2}E''(\zeta)(x_i - \bar{x})^2 \qquad (2.6.8)$$

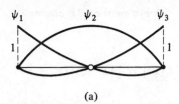

(a)

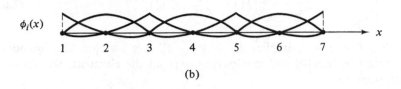

(b)

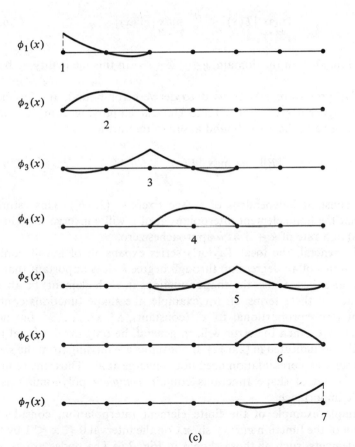

(c)

FIGURE 2.5 (a) A three-node element with quadratic shape functions; (b) a mesh consisting of three quadratic elements and (c) the basis functions generated by these elements.

whence

$$|E(\bar{x})| = \tfrac{1}{2}|E''(\zeta)|(x_i - \bar{x})^2 \qquad (2.6.9)$$

Since $x_{i+1} - x_i = h$, then $|x_i - \bar{x}| \le h/2$ in (2.6.6) and we have the error bound

$$|E(\bar{x})| \le \frac{h^2}{8}|E''(\zeta)| \qquad (2.6.10)$$

Finally, $E = g - g_h$ implies $E'' = g'' - g_h'' = g''$ within Ω_e. Introducing this result in (2.6.10) and maximizing over all the elements, we obtain the final estimate

$$\max_{0 \le x \le l} |E(x)| \le \frac{h^2}{8} \max_{0 \le x \le l} |g''(x)| \qquad (2.6.11)$$

Since g'' is bounded on the domain, $g'' \le C \le \infty$ in this inequality, C being a constant.

A similar procedure can be used to derive error bounds for Lagrange elements of higher degree. For a finite element employing complete polynomials of degree k, the error bound assumes the form

$$\|E\|_\infty = \max_{0 \le x \le l} |E(x)| \le Ch^{k+1} \qquad (2.6.12)$$

C being a constant independent of h (see Exercise (2.6.6)). This estimate indicates that the finite element interpolant g_h of g will converge to g (in the $\|\cdot\|_\infty$-norm) at a rate of $k + 1$ as h approaches zero.

Since, in general, the local Taylor's series expansion of g will contain polynomial terms of all degrees up through degree k, it is important that the interpolant (and, hence, also the shape functions in each element) be able to represent each of these terms. If, for example, the shape functions contain independent terms proportional to x^0 (constant), $x^2, x^3, \ldots, x^k$ but none proportional to x^1, then the error will, in general, be only proportional to h instead of h^{k+1} as indicated in (2.6.12). If constants are missing from the shape functions, then the representation need not converge at all. Thus, the requirement that the set of shape functions contain *complete* polynomials is of considerable importance.

As a simple example of the finite element interpolation, consider an interpolation of the function $g(x) = \sin \pi x$ on the interval $0 \le x \le 1$ by two quadratic elements, such as those shown in Fig. 2.6. The nodes are at $x = 0.0, 0.25, 0.50, 0.75,$ and 1.0, and the values of g at these nodes are $0.0, 0.707,$

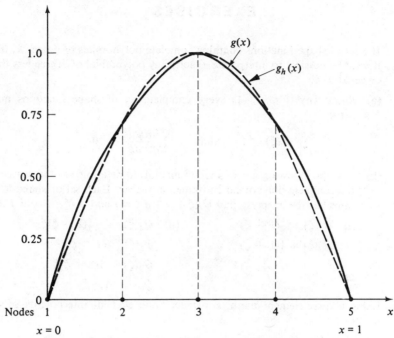

FIGURE 2.6 *Interpolation of* g(x) = *sin* πx *using two quadratic elements.*

1.0, 0.707, and 0.0, so that the finite element interpolant is

$$g_h(x) = 0.707\phi_2(x) + \phi_3(x) + 0.707\phi_4(x) \qquad (2.6.13)$$

To obtain an estimate of the interpolation error, note that $\max_{0 \le x \le 1} |g''(x)| = \pi^2$, so that $|g(x) - g_h(x)| \le ch^3$, where $c = \pi^2/48$.

A final comment of considerable importance should be made. In arriving at the error bound (2.6.12), we assumed that the given function g is so smooth that it has continuous derivatives of order $\le k$. Suppose that it does not! Assume, to the contrary, that g has continuous derivatives of only order s where $0 < s < k$. Then, *no matter how large the degree k of our interpolation polynomial g_h, only its first s terms may be effective in approximating g.* Then, instead of (2.6.12), we have

$$\max_{0 \le x \le l} |g(x) - g_h(x)| \le ch^s \qquad (2.6.14)$$

and the accuracy of our approximation, being independent of k, cannot be increased by increasing the degree of the polynomials defining ψ_i^e. We can, of course, improve the accuracy by reducing h as long as $s > 0$.

EXERCISES

2.6.1 If a set of shape functions contains complete polynomials of degree k, then it must be possible to interpolate exactly any polynomial of degree less than or equal to k.

(a) Show why (for $k \geq 1$) every complete set of shape functions must satisfy

$$\sum_{i=1}^{k+1} \hat{\psi}_i(\xi) = 1 \quad \text{and} \quad \sum_{i=1}^{k+1} \frac{d\hat{\psi}_i(\xi)}{d\xi} = 0$$

(b) Test the following sets of shape functions to determine k, the degree of the complete polynomial contained in the set. Each set of shape functions has the property that $\hat{\psi}_i(\xi_j) = 1$ if $i = j$ and $\hat{\psi}_i(\xi_j) = 0$ if $i \neq j$.

(i) $\hat{\psi}_1(\xi) = \frac{1}{4}(1 - \xi)^2$ (ii) $\hat{\psi}_1(\xi) = -\frac{1}{2}(1 - \xi)\xi$

$\quad\;\; \hat{\psi}_2(\xi) = \frac{1}{4}(1 + \xi)^2$ $\qquad\;\; \hat{\psi}_2(\xi) = (1 - \xi^2)^2$

$\quad\;\; \xi_1 = -1, \;\; \xi_2 = 1$ $\qquad\;\; \hat{\psi}_3(\xi) = \frac{1}{2}(1 + \xi)\xi$

$\qquad\qquad\qquad\qquad\qquad\qquad\;\; \xi_1 = -1, \;\; \xi_2 = 0, \;\; \xi_3 = 1$

2.6.2 Using a three-element mesh, construct finite element interpolants of the functions

$$f(x) = x + x^2 \quad \text{and} \quad g(x) = \cos \pi x$$

with $0 \leq x \leq 1$, using:

(a) Linear shape functions.

(b) Quadratic Lagrange shape functions.

Then compute error bounds in each of these approximations.

2.6.3 Compute the maximum error in the interpolant $g_h(x)$ in (2.6.13) of $\sin \pi x$ and compare it with the estimated error given in the text.

2.6.4 Derive explicit equations for cubic Lagrange shape functions and sketch them for a typical finite element. Illustrate the form of the basis functions produced by such shape functions for a mesh consisting of three elements.

2.6.5 It is possible to represent a given function $g = g(x)$ over an element by interpolating its values and the values of its derivatives at the endpoints of the element.

(a) Show that cubic shape functions can be used for this purpose.

(b) Sketch these shape functions for a typical finite element.

[*REMARK:* The shape functions produced in this way are called Hermite shape functions. They are discussed in some detail in Chapter 6.]

2.6.6 The error for $(k + 1)$-point Lagrange interpolation of a smooth function g is

$$E(x) = \frac{p_{k+1}(x)}{(k + 1)!} \frac{d^{k+1}g(\zeta)}{dx^{k+1}}, \quad x_1 \leq \zeta \leq x_{k+1}$$

where $\{x_i\}$ are the interpolation points and

$$p_{k+1}(x) = (x - x_1)(x - x_2) \ldots (x - x_{k+1}).$$

Using this result on the master element $-1 \leq \xi \leq 1$ and the map from $-1 \leq \xi \leq 1$ to $x \in \Omega_e$, show that

$$\max_{x \in \Omega_e} |E(x)| \leq \frac{h^{k+1}}{2^{k+1}(k+1)!} \max_{x \in \Omega_e} \left| \frac{d^{k+1}g(x)}{dx^{k+1}} \right|.$$

2.7 FINITE ELEMENT APPROXIMATION

At this point in our study, we have accumulated sufficient information to complete a detailed finite element analysis of second-order two-point boundary-value problems. In this section, we describe all of the steps through which (2.3.5) and its Galerkin approximation (2.4.4) can be used as a basis for the analysis of quite general two-point boundary-value problems by the finite element method. The procedure is outlined as follows.

2.7.1 Partitioning Ω and Selection of Shape Functions

We begin by partitioning Ω into a number of finite elements Ω_e of length h_e (thus $\sum_e h_e = l$). We will, for definiteness, assume that the domain of the exact solution of our problem is composed of the four natural smooth subdomains indicated in Fig. 2.2. Suppose that the concentrated source term in f shown in this figure is located at the coordinate $x = \bar{x}$ ($\bar{x} = x_2$ in Fig. 2.2). Then the flux $\sigma = -ku'$ will experience a jump $[\![\sigma]\!] = \hat{f}$ at $\bar{x}$. However, our element shape functions will always have the property that their derivatives are continuous within each element and, therefore, they cannot accommodate a jump such as this. *For this reason, we will always construct our mesh so that a nodal point is located at all points of discontinuity of the data.* Then terms such as $\hat{f}v_h(\bar{x})$ representing prescribed jumps will never enter the local equations which characterize the approximation over individual elements. These terms enter the analysis when the contributions of individual elements are summed.

With such a mesh in mind, we now focus our attention on a typical element Ω_e and consider possible choices of shape functions ψ_i^e. We have at our disposal any member of the Lagrange families of shape functions described in the preceding section. We could, for instance, use linear, quadratic, or cubic shape functions or, for that matter, shape functions consisting of polynomials of any degree k. Equation (2.6.12) suggests that the higher the degree k, the better accuracy we can expect if the solution u is sufficiently smooth. However, there is a price to pay for such accuracy.

As k increases, the bandwidth of the resulting stiffness matrix increases and thus, in general, so does the computational effort required in solving the final system of equations. For this reason (and some others*), it is rare that shape functions containing polynomials of degree higher than $k = 2$ or $k = 3$ are used in applications. We shall generally use linear or quadratic shape functions in the analysis of one-dimensional problems.

2.7.2 Calculation of Element Matrices and Equations

Having selected an appropriate set of shape functions, we now come to a crucial step in the analysis—the calculation of local approximations of the problem over each element. It is this aspect of finite element methods which leads some to say that "the problem is stated on each element" and the element statements are "patched together to form the final approximation of the problem."

To see how this is done, note that in the actual problem, for any smooth subdomain Ω_i between points s_1 and s_2, we have, for all admissible v,

$$\int_{s_1}^{s_2} (ku'v' + cu'v + buv)\, dx = \int_{s_1}^{s_2} \bar{f}v\, dx + \sigma(s_1)v(s_1) - \sigma(s_2)v(s_2) \qquad (2.7.1)$$

where $\sigma(s_i)$ is the flux at points s_i, $i = 1, 2$. Again note that the fluxes $\sigma(s_i)$ appear as given natural boundary data in the right-hand side of this equation.

Now let us consider a typical element Ω_e in the finite element mesh with endpoints s_1^e and s_2^e. Using (2.7.1), we formulate a variational statement of our problem for this single element, independent of whatever boundary conditions might be actually imposed at $x = 0$ and $x = l$. Thus, over each element we have a variational boundary-value problem of the form

$$\int_{s_1^e}^{s_2^e} (ku_h^{e\prime}v_h^{e\prime} + cu_h^{e\prime}v_h^e + bu_h^e v_h^e)\, dx$$
$$= \int_{s_1^e}^{s_2^e} \bar{f}v_h^e\, dx + \sigma(s_1^e)v_h^e(s_1^e) - \sigma(s_2^e)v_h^e(s_2^e) \qquad (2.7.2)$$

where u_h^e and v_h^e represent restrictions of u_h and v_h to Ω_e. It is important to realize that the $\sigma(s_i^e)$ are the true values of the flux at s_i^e and not their approximations; as explained earlier, the quantities $\sigma(s_i^e)$ correspond to natural boundary conditions at the endpoints of Ω_e. Note also that, unlike (2.4.5), no point-source terms such as $\hat{f}v_h^e(\bar{x})$ appear in (2.7.2) because of our decision to locate end nodal points of elements at these points.

We remark that we have chosen to use a global x-coordinate system in (2.7.2) only in order to clarify how the contributions from each element are

---

* Recall the comments in Section 2.6 leading up to (2.6.14).

actually summed together in generating the final stiffness and load matrices. In actual computations, these contributions are generally calculated for a master (reference) element in terms of the normalized local coordinate ξ in (2.6.1) and are then transformed to the appropriate coordinates for each element in the mesh. We describe this simple process in Chapter 3.

Note that u_h^e in (2.7.2) is of the form

$$u_h^e(x) = \sum_{j=1}^{N_e} u_j^e \psi_j^e(x) \tag{2.7.3}$$

where N_e is the number of nodes in Ω_e, ψ_j^e are the shape functions for this element, and u_j^e is the value of u_h^e at node x_j^e of the element,

$$u_j^e = u_h^e(x_j^e), \qquad j = 1, 2, \ldots, N_e \tag{2.7.4}$$

Upon substituting (2.7.3) into (2.7.2) and taking $v_h^e = \psi_i^e$, we arrive at a system of linear equations of the form

$$\sum_{j=1}^{N_e} k_{ij}^e u_j^e = f_i^e + \sigma(s_1^e)\psi_i^e(s_1^e) - \sigma(s_2^e)\psi_i^e(s_2^e), \qquad i = 1, 2, \ldots, N_e \tag{2.7.5}$$

In (2.7.5), k_{ij}^e are the entries in the element stiffness matrix and f_i^e are the components of the element load vector for Ω_e:

$$
\begin{aligned}
k_{ij}^e &= \int_{s_1^e}^{s_2^e} (k\psi_i^{e\prime}\psi_j^{e\prime} + c\psi_i^{e\prime}\psi_j^e + b\psi_i^e\psi_j^e)\, dx, \\
&\qquad\qquad\qquad\qquad\qquad\qquad i, j = 1, 2, \ldots, N_e \tag{2.7.6} \\
f_i^e &= \int_{s_1^e}^{s_2^e} f\psi_i^e\, dx,
\end{aligned}
$$

In actual finite element calculations, the integrals in (2.7.6) are rarely evaluated in closed form. Instead, the entries k_{ij}^e are generally computed using numerical integration rules,* which are of sufficient accuracy. Also, it is common practice to calculate f_i^e using the *interpolant* of f rather than f itself; for example, if $\bar{f}$ is the continuous part of f (excluding point sources) and if

$$f_h^e(x) = \sum_{i=1}^{N_e} \bar{f}(x_i^e)\psi_i^e(x)$$

then, instead of the formula in (2.7.6), we use

$$f_i^e = \int_{s_1^e}^{s_2^e} f_h^e\psi_i^e\, dx \tag{2.7.7}$$

---

* A discussion of numerical integration is given in Chapter 3.

In this way we can define the data f in our approximation by specifying its values at the nodal points.

2.7.3 Element Assembly

Having calculated the matrices and equations describing our approximation over each finite element, the next step in our analysis is to *assemble* the equations describing the approximation on the entire mesh by adding up the contributions to these equations furnished by each element.

To fix ideas, consider the special case in which linear shape functions of the form in (2.6.4) are used. Each element then has two nodes, and therefore there are two equations per element of the following form:

$$\left. \begin{array}{l} k_{11}^e u_1^e + k_{12}^e u_2^e = f_1^e + \sigma(s_1^e) \\ k_{21}^e u_1^e + k_{22}^e u_2^e = f_2^e - \sigma(s_2^e) \end{array} \right\} \tag{2.7.8}$$

Here the subscripts 1 and 2 are labels of the endpoint nodes on a typical element and $\sigma(s_1^e)$ and $\sigma(s_2^e)$ represent the actual values of the flux $\sigma = -ku'$ at the nodes. Of course, these subscripts are to be relabeled upon assembling the elements so as to coincide with appropriate node numbers $1, 2, 3, \ldots, N$ in the final mesh. For example, if the element is to fit between nodes 6 and 7 in a mesh, u_1^e in (2.7.8) is actually u_6, u_2^e is u_7, $\sigma(s_1^e)$ is the value of $-ku'$ as the node at x_6 is approached from the right, and $\sigma(s_2^e)$ is $-ku'$ as x_7 is approached from the left.

We now *assemble* the equations describing the entire collection of elements comprising our mesh by sweeping through all elements, one at a time, and using (2.7.8) to calculate the contributions of each of them. Consider, for example, a mesh containing $N - 1$ elements and N nodes, numbered consecutively $1, 2, \ldots, N$, as shown in Fig. 2.7. This means that there results N equations in N degrees of freedom describing the assembled system of elements, and we must allocate space in the computer for a system of this size. Thus, we anticipate calculating an $N \times N$ stiffness matrix $\mathbf{K} = [K_{ij}]$ and an $N \times 1$ load vector $\mathbf{F} = \{F_i\}$, $i, j = 1, 2, \ldots, N$.

We initiate the assembly process by setting $K_{ij} = 0$ and $F_i = 0$. For element Ω_1, between nodes 1 and 2, (2.7.8) yields the equations

$$k_{11}^1 u_1 + k_{12}^1 u_2 = f_1^1 + \sigma(0)$$
$$k_{21}^1 u_1 + k_{22}^1 u_2 = f_2^1 - \sigma(x_1^-)$$

where $\sigma(0) = \sigma(0^+)$ is the actual flux at node 1 as this node is approached

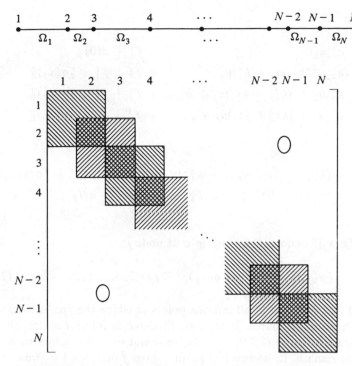

FIGURE 2.7 *A finite-element mesh with N nodes and N-1 elements and the assembled stiffness matrix with the shaded blocks of entries representing the contributions of each element; the symbols 0 represent the fact that outside the diagonal blocks all entries are zero.*

from the right and $\sigma(x_1^-)$ is the flux at the node at x_1 as this node is approached from the left. These equations are added into the first and second rows of the $N \times N$ system describing the entire mesh.

We next go to element Ω_2. Since it lies between nodes 2 and 3, its contributions, calculated using (2.7.8), are added to the equations in rows 2 and 3. Since two elements and three nodes have been "activated," we now have the three equations,

$$k_{11}^1 u_1 + k_{12}^1 u_2 \qquad\qquad = f_1^1 + \sigma(0)$$
$$k_{21}^1 u_1 + (k_{22}^1 + k_{11}^2)u_2 + k_{12}^2 u_3 = f_2^1 + f_1^2 - \sigma(x_1^-) + \sigma(x_1^+)$$
$$k_{21}^2 u_2 + k_{22}^2 u_3 = f_2^2 - \sigma(x_2^-)$$

Continuing this process through the entire system of N elements, we arrive at the system

$$
\begin{aligned}
k_{11}^1 u_1 + k_{12}^1 u_2 \qquad\qquad\qquad &= f_1^1 + \sigma(0) \\
k_{21}^1 u_1 + (k_{22}^1 + k_{11}^2)u_2 + k_{12}^2 u_3 \qquad\qquad &= f_2^1 + f_1^2 + [\![\sigma(x_1)]\!] \\
k_{21}^2 u_2 + (k_{22}^2 + k_{11}^3)u_3 + k_{12}^3 u_4 &= f_2^2 + f_1^3 + [\![\sigma(x_2)]\!] \\
k_{21}^3 u_3 + (k_{22}^3 + k_{11}^4)u_4 + \ldots \quad &= f_2^3 + f_1^4 + [\![\sigma(x_3)]\!] \\
& \vdots \\
k_{21}^{N-2} u_{N-2} + (k_{22}^{N-2} + k_{11}^{N-1})u_{N-1} + k_{12}^{N-1} u_N &= f_2^{N-2} + f_1^{N-1} + [\![\sigma(x_{N-1})]\!] \\
k_{21}^{N-1} u_{N-1} + k_{22}^{N-1} u_N &= f_2^{N-1} - \sigma(l)
\end{aligned}
\right\}
$$

$$(2.7.9)$$

wherein $[\![\sigma(x_j)]\!]$ denotes the jump in σ at node j:

$$[\![\sigma(x_j)]\!] = \sigma(x_j^+) - \sigma(x_j^-), \qquad j = 2, 3, \ldots, N - 1 \qquad (2.7.10)$$

Recall that $[\![\sigma]\!] = 0$ at all interior points at which the flux is continuous. If there are no point sources in the data f located at interior nodes, all of the interior jump terms in (2.7.9) must be zero and only the values of σ at the boundaries remain. If, however, a point source $\hat{f}_j \delta(x - x_j)$ is prescribed at any interioi node x_j, then we must set

$$[\![\sigma(x_j)]\!] = \hat{f}_j$$

in (2.7.9).

Let us assume that a point source $\hat{f}\delta(x - \bar{x})$ is located at node 3: $\bar{x} = x_3$. Then the linear system of equations for the entire mesh assumes the form

$$
\begin{bmatrix}
\tilde{K}_{11} & K_{12} & 0 & 0 & \cdots & 0 & 0 \\
K_{21} & K_{22} & K_{23} & 0 & \cdots & 0 & 0 \\
0 & K_{32} & K_{33} & K_{34} & \cdots & 0 & 0 \\
0 & 0 & K_{43} & K_{44} & \cdots & 0 & 0 \\
\cdot & \cdot & \cdot & \cdot & \cdots & \cdot & \cdot \\
0 & 0 & 0 & 0 & \cdots & K_{N-1,N-1} & K_{N-1,N} \\
0 & 0 & 0 & 0 & \cdots & K_{N,N-1} & \tilde{K}_{NN}
\end{bmatrix}
\begin{bmatrix}
u_1 \\ u_2 \\ u_3 \\ u_4 \\ \vdots \\ u_{N-1} \\ u_N
\end{bmatrix}
=
\begin{bmatrix}
\tilde{F}_1 \\ F_2 \\ F_3 \\ F_4 \\ \vdots \\ F_{N-1} \\ \tilde{F}_N
\end{bmatrix}
$$

$$(2.7.11)$$

where the $N \times N$ coefficient matrix contains the entries

$$
\begin{bmatrix}
k_{11}^1 & k_{12}^1 & 0 & 0 & \cdots & 0 & 0 \\
k_{21}^1 & k_{22}^1 + k_{11}^2 & k_{12}^2 & 0 & \cdots & 0 & 0 \\
0 & k_{21}^2 & k_{22}^2 + k_{11}^3 & k_{12}^3 & \cdots & 0 & 0 \\
0 & 0 & k_{21}^3 & k_{22}^3 + k_{11}^4 & \cdots & 0 & 0 \\
\cdot & \cdot & \cdot & \cdot & & \cdot & \cdot \\
0 & 0 & 0 & 0 & \cdots & k_{22}^{N-2} + k_{11}^{N-1} & k_{12}^{N-1} \\
0 & 0 & 0 & 0 & \cdots & k_{21}^{N-1} & k_{22}^{N-1}
\end{bmatrix}
$$

$$(2.7.12)$$

and

$$
\begin{bmatrix}
\tilde{F}_1 \\
F_2 \\
F_3 \\
\cdot \\
\cdot \\
\cdot \\
F_{N-1} \\
\tilde{F}_N
\end{bmatrix}
=
\begin{bmatrix}
f_1^1 + \sigma(0) \\
f_2^1 + f_1^2 \\
f_2^2 + f_1^3 + \hat{f} \\
\cdot \\
\cdot \\
\cdot \\
f_2^{N-2} + f_1^{N-1} \\
f_2^{N-1} - \sigma(l)
\end{bmatrix}
;
\qquad (2.7.13)
$$

The stiffness coefficients $\tilde{K}_{11}$ and $\tilde{K}_{NN}$ and the load components $\tilde{F}_1$ and $\tilde{F}_N$ are not yet of the general form described in (2.4.4) and (2.4.5) because they do not contain the boundary contributions. We discuss modifications in these terms for various choices of boundary conditions subsequently.

Notice that the stiffness matrix in (2.7.11) is sparse (there are many zeros) and that the location of the element matrices, indicated in the bordered blocks in (2.7.11), overlap in rows and columns corresponding to shared nodes. Note also that if the coefficient c in (2.7.1) is not identically zero, the matrix will be unsymmetric.

2.7.4 Boundary Conditions

An extremely important feature of the development up to this point is that no boundary conditions have, as yet, been applied. Thus, (2.7.11) is applicable to a wide range of boundary conditions.

Consider, for instance, the following cases:

1. **General Natural Boundary Conditions:** These correspond to the general case (2.2.2) in which a linear combination of u and u' are prescribed:

$$\alpha_0 u'(0) + \beta_0 u(0) = \gamma_0, \qquad \alpha_l u'(l) + \beta_l u(l) = \gamma_l \qquad (2.7.14)$$

In our approximation of this case, we set

$$u_h'(0) = \frac{\gamma_0 - \beta_0 u_h(0)}{\alpha_0} \quad \text{and} \quad u_h'(l) = \frac{\gamma_l - \beta_l u_h(l)}{\alpha_l} \qquad (2.7.15)$$

where, of course, $u_h(0) = u_1$ and $u_h(l) = u_N$. Then (2.7.11) reduces to

$$
\begin{bmatrix}
\tilde{K}_{11} - \dfrac{k(0)\beta_0}{\alpha_0} & K_{12} & 0 & \cdots & 0 & 0 \\[2mm]
K_{21} & K_{22} & K_{23} & \cdots & & \\[1mm]
0 & K_{32} & K_{33} & \cdots & & \\[1mm]
\multicolumn{6}{c}{\cdots\cdots\cdots\cdots\cdots\cdots\cdots\cdots\cdots\cdots\cdots} \\[1mm]
0 & 0 & 0 & \cdots & K_{N-1,N-1} & K_{N-1,N} \\[2mm]
0 & 0 & 0 & \cdots & K_{N,N-1} & \tilde{K}_{NN} + \dfrac{k(l)\beta_l}{\alpha_l}
\end{bmatrix}
\times
$$

$$
\begin{bmatrix} u_1 \\ u_2 \\ u_3 \\ \vdots \\ u_{N-1} \\ u_N \end{bmatrix}
=
\begin{bmatrix} f_1^1 - \dfrac{k(0)\gamma_0}{\alpha_0} \\[2mm] F_2 \\ F_3 \\ \vdots \\ F_{N-1} \\ f_2^{N-1} + \dfrac{k(l)\gamma_l}{\alpha_l} \end{bmatrix}
\qquad (2.7.16)
$$

If the final $N \times N$ stiffness matrix in (2.7.16) is invertible, we can solve (2.7.15) for the unknown nodal values $u_1, u_2, \ldots, u_N$. Other features of the approximation, such as the approximate flux $\sigma_h = -ku_h'$, can then be easily evaluated.

2.   Dirichlet Boundary Conditions: Boundary conditions of the type

$$u(0) = \frac{\gamma_0}{\beta_0}, \qquad u(l) = \frac{\gamma_l}{\beta_l} \qquad (2.7.17)$$

follow as a special case of (2.7.14) when $\alpha_0 = \alpha_l = 0$. Essential boundary conditions of this form are usually called *Dirichlet* boundary conditions and the corresponding boundary-value problem is referred to as a *Dirichlet problem* for the function u.

In this case, $u_h(0) = u_1 = \gamma_0/\beta_0$ and $u_h(l) = u_N = \gamma_l/\beta_l$, so that only $N - 2$ unknown nodal values $u_2, u_3, \ldots, u_{N-1}$ remain. Then (2.7.12) reduces to the $(N - 2) \times (N - 2)$ system

$$\begin{bmatrix} K_{22} & K_{23} & 0 & \cdots & 0 & 0 \\ K_{32} & K_{33} & K_{34} & \cdots & 0 & 0 \\ 0 & K_{43} & K_{44} & \cdots & 0 & 0 \\ \multicolumn{6}{c}{\cdots\cdots\cdots\cdots\cdots\cdots\cdots\cdots\cdots} \\ 0 & 0 & 0 & \cdots & K_{N-1,N-2} & K_{N-1,N-1} \end{bmatrix} \begin{bmatrix} u_2 \\ u_3 \\ u_4 \\ \vdots \\ u_{N-1} \end{bmatrix}$$

$$= \begin{bmatrix} F_2 - \dfrac{K_{21}\gamma_0}{\beta_0} \\ F_3 \\ F_4 \\ \vdots \\ F_{N-1} - \dfrac{K_{N-1,N}\gamma_l}{\beta_l} \end{bmatrix} \qquad (2.7.18)$$

and the two auxiliary equations corresponding to nodes 1 and N,

$$\left. \begin{aligned} \tilde{K}_{11}\left(\frac{\gamma_0}{\beta_0}\right) + K_{12}u_2 &= f_1^1 + \sigma(0) \\ K_{N,N-1}u_{N-1} + \tilde{K}_{NN}\left(\frac{\gamma_l}{\beta_l}\right) &= f_2^{N-1} - \sigma(l) \end{aligned} \right\} \qquad (2.7.19)$$

The reduced stiffness matrix in (2.7.18) is nonsingular, so that (2.7.18) can be solved for the unknown nodal values $u_2, u_3, \ldots, u_{N-1}$. Of course, (2.7.19) also provides useful information. Once u_2 and u_{N-1} are known, the approximations of the endpoint fluxes can be computed directly using (2.7.19).

3. **Neumann Boundary Conditions:** When only the derivative of u is specified at each end, (2.7.14) reduces to

$$u'(0) = \frac{\gamma_0}{\alpha_0}, \qquad u'(l) = \frac{\gamma_l}{\alpha_l} \qquad (2.7.20)$$

whenever $\beta_0 = \beta_l = 0$. Natural boundary conditions of this type are called *Neumann* boundary conditions and the corresponding boundary-value problem is referred to as a *Neumann problem* for the function u.

Neumann problems frequently require some special considerations for certain forms of the governing differential equation. In particular, consider the case in which the coefficients $c = c(x)$ and $b = b(x)$ are identically zero so that the boundary-value problem becomes one of solving the differential equation

$$-(k(x)u'(x))' = f(x) \qquad (2.7.21)$$

on smooth subdomains, subject to the end conditions (2.7.20). In this case, the solution u is determined only to within an arbitrary constant c_0; that is, if $u = u(x)$ satisfies (2.7.20) and (2.7.21), then $u + c_0$ is also a solution. Because of the analogy of (2.7.20) with equations describing mechanical systems, the constant c_0 is sometimes referred to as a *rigid motion*, and this rigid motion must be specified if we are to obtain a unique solution to our problem. Moreover, the finite element approximation (2.7.11) of this Neumann problem will also contain an arbitrary rigid motion. Since solutions to (2.7.11) will then be non-unique, the stiffness matrix $\mathbf{K}$ in (2.7.12) will necessarily be singular.

The presence of a rigid motion in the solution to a Neumann problem leads to another consideration of fundamental importance: the data f, γ_0, α_0, γ_l, and α_l *cannot be specified* arbitrarily, they must be *compatible* in a sense we will now make clear. Since the variational form of this Neumann problem (with $c, b \equiv 0$) is to find $u \in H^1$ such that

$$\int_0^l ku'v' \, dx = \int_0^l \bar{f}v \, dx + \hat{f}v(\bar{x})$$
$$-k(0)\left(\frac{\gamma_0}{\alpha_0}\right)v(0) + k(l)\left(\frac{\gamma_l}{\alpha_l}\right)v(l) \qquad \text{for all } v \in H^1 \qquad (2.7.22)$$

and since $u = c_0 = $ constant is a solution, this equation must also hold for $u = c_0$ and the choice $v = 1$. Hence, the data must be such that

$$\int_0^l \bar{f} \, dx + \hat{f} - \frac{k(0)\gamma_0}{\alpha_0} + \frac{k(l)\gamma_l}{\alpha_l} = 0 \qquad (2.7.23)$$

The compatibility condition (2.7.23) is a *necessary* condition for the existence of a solution to (2.7.22). We remark that from a physical viewpoint, (2.7.23) is a global conservation law; it reflects the requirement that the flux σ be conserved over the entire body Ω.

For the discrete problem corresponding to (2.7.22), this compatibility condition assumes the form (see (2.7.13))

$$\tilde{F}_1 + \sum_{i=2}^{N-1} F_i + \tilde{F}_N = 0 \qquad (2.7.24)$$

To eliminate the rigid motion, we can specify the value u_j of u_h at any node j equal to an arbitrary constant c_0. For instance, setting $u_1 = c_0$ in (2.7.16) (with it understood that $c = b = 0$) yields the reduced system

$$
\begin{bmatrix}
K_{22} & K_{23} & 0 & \cdots & 0 & 0 \\
K_{32} & K_{33} & K_{34} & \cdots & 0 & 0 \\
0 & K_{43} & K_{44} & \cdots & 0 & 0 \\
\multicolumn{6}{c}{\cdots\cdots\cdots\cdots\cdots\cdots\cdots\cdots} \\
0 & 0 & 0 & \cdots & K_{N,N-1} & \tilde{K}_{NN}
\end{bmatrix}
\begin{bmatrix}
u_2 \\ u_3 \\ u_4 \\ \cdot \\ \cdot \\ \cdot \\ u_N
\end{bmatrix}
=
\begin{bmatrix}
F_2 - K_{21}c_0 \\ F_3 \\ F_4 \\ \cdot \\ \cdot \\ f_2^{N-1} + \dfrac{k(l)\gamma_l}{\alpha_l}
\end{bmatrix}
$$

$$(2.7.25)$$

and the equation

$$\tilde{K}_{11}c_0 + K_{12}u_2 = f_1^1 - \frac{k(0)\gamma_0}{\alpha_0} \qquad (2.7.26)$$

Equation (2.7.25) is uniquely solvable for $u_2, u_3, \ldots, u_N$ in terms of c_0. Frequently, we simply set $c_0 = 0$. Notice that we can solve for u_2 using either (2.7.25) or (2.7.26). It is a remarkable fact that the condition (2.7.24) guarantees that these two systems will be compatible: the value of u_2 obtained from (2.7.25) will be exactly the same as that of (2.7.26) whenever (2.7.24) holds.

4. **Mixed Boundary Conditions:**   When an essential boundary condition is applied at one boundary point and a natural boundary condition at the other, a *mixed boundary-value problem* for the function u is obtained. For example, one mixed problem is characterized by the end conditions

$$u(0) = \frac{\gamma_0}{\beta_0}, \qquad u'(l) = \frac{\gamma_l}{\alpha_l} \qquad (2.7.27)$$

and another by

$$\alpha_0 u'(0) + \beta_0 u(0) = \gamma_0, \qquad u(l) = \frac{\gamma_l}{\beta_l} \qquad (2.7.28)$$

Since at least one of these conditions specifies the value of u at an endpoint, the solution will contain no rigid motions.

Consider the case (2.7.27). Then $u_1 = \gamma_0/\beta_0$, $\sigma(l) = -k(l)\gamma_l/\alpha_l$, and (2.7.11) reduces to the system

$$\begin{bmatrix} K_{22} & K_{23} & \cdots & 0 & 0 \\ K_{32} & K_{33} & \cdots & 0 & 0 \\ \multicolumn{5}{c}{\cdots\cdots\cdots\cdots\cdots\cdots\cdots} \\ 0 & 0 & \cdots & K_{N,N-1} & \tilde{K}_{NN} \end{bmatrix} \begin{bmatrix} u_2 \\ u_3 \\ \cdot \\ \cdot \\ u_N \end{bmatrix} = \begin{bmatrix} F_2 - \dfrac{K_{21}\gamma_0}{\beta_0} \\ F_3 \\ \cdot \\ \cdot \\ f_2^{N-1} + \dfrac{k(l)\gamma_l}{\alpha_l} \end{bmatrix}$$

$$(2.7.29)$$

and the equation

$$\tilde{K}_{11}\left(\frac{\gamma_0}{\beta_0}\right) + K_{12}u_2 = f_1^1 - \sigma(0) \qquad (2.7.30)$$

We solve (2.7.29) for $u_2, u_3, \ldots, u_N$ and use (2.7.30) to obtain an approximation $\sigma_h(0)$ of $\sigma(0)$, if desired. Conditions (2.7.28) are handled in a similar fashion. We leave these details as an exercise.

2.7.5 Error Estimates

Suppose that the actual solution u of our boundary-value problem has the property that its derivatives of order s are square-integrable on Ω, but those of order $s + 1$ and higher are not, s being an integer greater than unity. Further, suppose that we use shape functions that contain complete polynomials of degree $\leq k$ and a uniform mesh of elements of equal length h. Then the approximation error, measured in the H^1-norm of (2.3.11), can be shown to satisfy the asymptotic error estimate

$$\|u - u_h\|_1 \leq Ch^\mu \qquad (2.7.31)$$

where C is a constant independent of h and

$$\mu = \min(k, s) \qquad (2.7.32)$$

That is, the rate of convergence is k if $k < s$ or s if $s < k$.

The behavior of the error with respect to the mean-square norm is an order better,

$$\|u - u_h\|_0 \leq C_1 h^{\mu+1} \tag{2.7.33}$$

with C_1 a constant.

The estimates (2.7.31) and (2.7.33) indicate that when the solution u is regular (i.e., $s > k$), then an improvement in the rate of convergence is obtained by increasing the degree k of the polynomial used in the approximation. However, for $s < k$, the rate of convergence is independent of k and no improvement is observed by increasing k.

This completes our description of the finite element analysis of two-point boundary-value problems of the type in (2.3.5). What remains to be done is to implement the procedure described above by developing a finite element computer program. The general features of such a program are clearly suggested by the steps just described. We shall furnish more detailed information on this important aspect of finite element analysis in Chapter 3.

EXERCISES

2.7.1 Consider the boundary-value problems defined by the differential equation

$$-u''(x) + b_0 u(x) = 10\,\delta(x - 1), \qquad 0 < x < 2$$

b_0 being a constant, and the following sets of boundary conditions:

(i) $u(0) = 1, u(2) = 3$

(ii) $u'(0) = 2, u'(2) = g_0$ $(g_0 = \text{constant})$

(iii) $u'(0) + u(0) = 1, u(2) = 1$

(a) Using four elements of equal length and piecewise-linear basis functions, compute the global stiffness matrix and load vectors for this general class of problems for the cases in which $b_0 = 1$ and $b_0 = 0$. Give numerical values of all entries.

(b) Develop the reduced (nonsingular) equations for problems (i) and (iii).

(c) Consider conditions (ii) and $b_0 = 0$. What value must the constant g_0 be given?

(d) Develop the reduced equations for (ii) with $b_0 = 0$ and g_0 the constant determined in part (c).

2.7.2 Consider the Neumann problem

$$-u''(x) = f(x), \qquad 0 < x < 1$$
$$u'(0) = 1, \qquad u'(1) = 2$$

(a) Develop the linear system of equations describing an approximation of this problem using only two finite elements and linear shape functions ($N = 3$). The actual numerical values of K_{ij} and F_i need not be calculated.

(b) Use the arguments leading up to (2.7.25) and (2.7.26) to reduce this system to an invertible 2×2 system of equations and an auxiliary equation for the value of u_h at a specified node. What are the conditions on f and F_1, F_2, F_3?

(c) Show that the 2×2 system and the auxiliary equation derived in (b) are compatible (i.e., yield the same answer).

[*HINT:* Observe that $\displaystyle\sum_{i,j=1}^{3} K_{ij}v_j = \sum_{i,j=1}^{3} K_{ij}w_i v_j = \int_0^1 k w_h' v_h'\, dx = 0$ for $w_h = 1$ and all admissible v_h.]

3

DEVELOPMENT OF A FINITE
ELEMENT PROGRAM

3.1 COMPUTER IMPLEMENTATION
OF THE FINITE ELEMENT METHOD

The success of the finite element method is due, in no small degree, to the
ease with which the method can be implemented in the form of *general-
purpose* computer programs.* The term "general purpose," applied to finite
element programs, means that any boundary-value problem in a certain class
can be completely specified by means of the input data alone (i.e., without
requiring changes in the program). The broader the class of problems, of
course, the more general is the purpose of the program—considerable generality is attainable with the finite element method.

In this chapter, we discuss the development of a finite element code for
the solution of a fairly general class of two-point boundary-value problems.
Some fundamental considerations of code design are presented and the
overall structure of the code is outlined. Some details of coding are given,
but many are left as exercises for the reader. In fact, this chapter is designed
to allow the reader to complete the development of the code, including the
debugging, verification, and documentation phases and to use the code to
study the behavior of the finite element method through numerical experiments and simulated applications. Our aim is neither great generality nor
sophistication in coding—it is, rather, to convey the basic concepts and

---

* The words "program" and "code" are used interchangeably in this volume.

techniques by means of which the finite element method becomes a useful tool.

As viewed by the user, a finite element code consists of the three functional units shown in Fig. 3.1. Because the finite element modeling of each

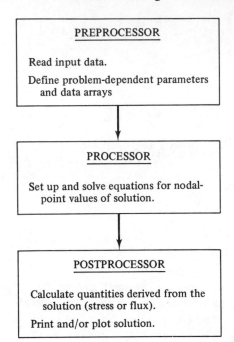

FIGURE 3.1 *Functional units of a finite element code.*

boundary-value problem is defined by the input data, there is typically a large amount of data required by the program. Significant portions of the data can be generated internally when regular meshes are to be used. The reading and generation of the arrays of data that define the finite element model of the boundary-value problem are performed by the *preprocessing* unit of the program. The preprocessor may consist of a few statements, a single subroutine, or even a complex program linked to the other units through disk or tape files.

The actual finite element calculations, as described in Section 3.3, are carried out in the processor, which is usually considered to be of central importance. In this part of the program element matrices and vectors are calculated and assembled, boundary conditions are imposed, and the global set of equations is solved for the nodal-point values of the solution.

The postprocessing of the finite element solution can vary in complexity from simply printing the nodal-point values of the solution to calculating the flux ($\sigma = -ku'$), contouring and plotting selected results, and comparing the solution with allowable values.

3.2 DESCRIPTION OF CODE1

We now describe some general features of a finite element code, called CODE1, which is designed to perform a finite element analysis of a class of two-point boundary-value problems. A discussion of details on the construction of CODE1 and the various subroutines it contains comprises the remainder of this chapter. We are concerned with a relatively general class of one-dimensional problems characterized by a differential equation and boundary conditions of the following form:

Differential equation:

$$-\frac{d}{dx}\left(k(x)\frac{du}{dx}\right) + c(x)\frac{du(x)}{dx} + b(x)u(x) = f(x), \qquad a < x < b \qquad (3.2.1)$$

Boundary conditions:

$$
\left.
\begin{array}{ll}
u(a) = u_a & u(b) = u_b \\[2mm]
k(a)\dfrac{du(a)}{dx} = \sigma_a & -k(b)\dfrac{du(b)}{dx} = \sigma_b \\[2mm]
k(a)\dfrac{du(a)}{dx} = p_a[u(a) - u_a] & -k(b)\dfrac{du(b)}{dx} = p_b[u(b) - u_b]
\end{array}
\right\} \qquad (3.2.2)
$$

In the coding presented in the text, the coefficients (called material properties) k, c, and b, and the source f are constant within each of up to five subregions (materials). More general cases are dealt with in the exercises. The right-hand side (load), in addition to the regular part, which is treated as a material property, may have up to five individual point loads (Dirac deltas). The extent of the region and each subregion, the values of material properties and loads, and the type of boundary conditions as well as boundary data are defined by data input to the program.

The finite element modeling of the problem consists of variable-length finite elements of linear, quadratic, or cubic degree, which are defined by input data. The output from the program consists of the finite element solution $u_h(x)$ and flux $\sigma_h(x) = -k(x)u_h'(x)$ at user-selected points. A user-supplied subroutine for evaluating the exact solution u and flux σ can be used to evaluate errors at output points and/or calculate norms of the error in the finite element solution.

The flowchart in Fig. 3.2 shows each subroutine in CODE1 and its relation to other routines. Routine names are underlined and a brief description of their function is given at their first appearance. An asterisk following a routine name indicates that some or all of the coding for the routine is given in the text.

A very important factor in the design and development of any finite

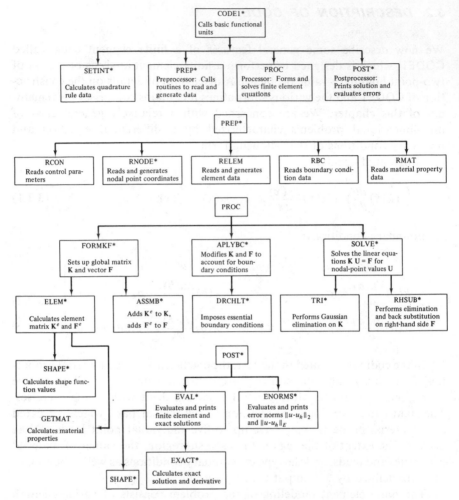

FIGURE 3.2   *Flow charts of components of CODE1.*

element code is the management of the large amounts of data required by the processor. One reasonable classification of these data considers nodal point data, element data, material property data, and boundary condition data as comprising the base of data to be managed. In addition to these lists of data, there is a set of parameters, which we will refer to as control data, that serves to define the data management and computational procedures. Finally, there are the arrays of the global stiffness matrix and load vector used in the solution of the finite element equations. In applications codes, many of these data sets exceed the amount of core storage available on the user's computer. In such cases, schemes of various degrees of sophistication are used to make the data available as needed. The data in CODE1 are organized into several labeled COMMON blocks, whose content is summarized in Fig. 3.3. As a general rule, the data are defined in routines called by the preprocessor PREP and used in the remainder of the code. The description in the text of each routine indicates which COMMON blocks are to be included in the routine.

The main program serves only to call the routine SETINT that initializes numerical integration data (see Section 3.3.1) and the three main units of the code shown in Fig. 3.1. The coding of the main program is as follows:

```
      PROGRAM CODE1(INPUT,OUTPUT,TAPE5=INPUT,TAPE6=OUTPUT)
      COMMON/CCON/TITLE(20),NNODE,NELEM,NMAT,NPOINT
      COMMON/CELEM/KIND(99),MAT(99),NODES(4,99),NINT(99)
      COMMON/CNODE/X(100),U(100)
      COMMON/CMATL/PROP(4,5)
      COMMON/CBC/KBC(2),VBC(2,2),KPOINT(5),POINT(5)
      COMMON/CMATRX/GK(100,100),GF(100)
      COMMON/CINT/XI(4,4),W(4,4)
      CALL SETINT
    1 CALL PREP
      CALL PROC
      CALL POST
      GO TO 1
      END
```

3.3 ELEMENT CALCULATIONS

In the heart of any finite element code lies the calculation and assembly of the element matrices. Our description of the development of this prototype code begins with a detailed description of the subroutines that perform the element calculations. These calculations consist of the integrations indicated in the variational statement of the boundary-value problem and are carried out in subroutine ELEM in conjunction with routines SETINT, SHAPE, and GETMAT.

Since all of the coefficients and basis functions that form the integrands in the variational statement are smooth within each element, this integration can be carried out without any concern for smoothness (see Section 2.7). For problems of the class considered here (one-dimensional elements, polynomial basis functions, constant coefficients) exact integration could be readily performed. Since, in more interesting applications, these conditions rarely hold, we base the coding of the element calculations on numerical integration. The technique is general, powerful, and almost universally used in applications of finite element methods.

CONTROL PARAMETERS

COMMON/CCON/TITLE(20), NNODE, NELEM, NMAT, NPOINT

TITLE — alphanumeric array containing problem title
NNODE — number of nodal points in problem
NELEM — number of elements in problem
NMAT — number of materials in problem
NPOINT — number of point loads in problem

NODAL–POINT DATA

COMMON/CNODE/X(100), U(100)

X — coordinates of nodal points
U — finite element solution at nodal points

ELEMENT DATA

COMMON/CELEM/KIND(99), MAT(99), NODES(4,99), NINT(99)

KIND — degree of polynomial in element shape function
MAT — number of material type of element
NODES — node numbers of nodes in element
NINT — number of integration points used in element calculations

MATERIAL DATA

COMMON/CMATL/PROP(4,5)

PROP — coefficients k, c, b, and f for the material

BOUNDARY–CONDITION DATA

COMMON/CBC/KBC(2), VBC(2,2), KPOINT(5), POINT(5)

KBC — boundary condition flag indicating type of boundary condition at each end
VBC — values of boundary-condition data
KPOINT — node number of point loads
POINT — value of point loads

MATRIX STORAGE

COMMON/CMATRX/GK(100, 100), GF(100)

GK — global **K** matrix
GF — global **F** vector

NUMERICAL QUADRATURE DATA

COMMON/CINT/XI(4, 4), W(4, 4)

XI — location of integration points for rules of order 1–4
W — weights for integration points for rules of order 1–4

FIGURE 3.3 Data arrays in CODE1.

92

The contribution of an element with endpoints x_1 and x_2 and N nodes is

$$\int_{x_1}^{x_2} (ku_h'v_h' + cu_h'v_h + bu_hv_h - fv_h)\, dx \tag{3.3.1}$$

Substitution of the shape function expansions for $u_h(x)$ and $v_h(x)$ yields the explicit expressions for the element stiffness matrix $\mathbf{k}^e$ and the element load vector $\mathbf{f}^e$ (recall (2.7.6)):

$$\left.\begin{aligned}
k_{ij}^e &= \int_{x_1}^{x_2} [k(x)\psi_i^{e\prime}(x)\psi_j^{e\prime}(x) + c(x)\psi_i^e(x)\psi_j^{e\prime}(x) + b(x)\psi_i^e(x)\psi_j^e(x)]\, dx \\
f_i^e &= \int_{x_1}^{x_2} \bar{f}(x)\psi_i^e(x)\, dx, \qquad i,j = 1, 2, \ldots, N
\end{aligned}\right\} \tag{3.3.2}$$

Clearly, to compute each of the quantities in (3.3.2), we must evaluate integrals of the form

$$\mathcal{I} = \int_{x_1}^{x_2} g(x)\, dx \tag{3.3.3}$$

The numerical evaluation of the integral $\mathcal{I}$ in (3.3.3) is accomplished by choosing a *quadrature formula* of the form

$$\mathcal{I} \doteq I = \sum_{l=1}^{N_I} g(\bar{x}_l)w_l \tag{3.3.4}$$

where $\bar{x}_l$ are the integration points in the interval $x_1 \le x \le x_2$; the w_l are numbers, called *weights*; and N_I is the *order* of the formula. Again, we view these calculations as being performed on a master element $\hat{\Omega}$ defined by the interval $-1 \le \xi \le 1$, so that limits of integration are -1 and 1. Specifically, if we choose as the transformation from ξ to x the relation

$$x = x_1 + \tfrac{1}{2}(x_2 - x_1)(1 + \xi) \tag{3.3.5}$$

then we have $x = x_1$ when $\xi = -1$, $x = x_2$ when $\xi = 1$, and $dx = (dx/d\xi)\, d\xi = \tfrac{1}{2}(x_2 - x_1)\, d\xi$. The form of the quadrature formula (3.3.4) now becomes

$$I = \tfrac{1}{2}(x_2 - x_1) \sum_{l=1}^{N_I} g(x(\xi_l))w_l \tag{3.3.6}$$

The values of ξ_l, and (the new) w_l for $l = 1, \ldots, N_I$ completely determine the numerical integration rule.

In most finite element calculations, *Gaussian quadrature* rules are used.

We note that the Gauss rule of order N_l will integrate exactly polynomials of degree $2N_l - 1$.

3.3.1 Integration Routine—SETINT

The integrations indicated by (3.3.2) and (3.3.3) are carried out in CODE1 using Gaussian quadrature of order 1, 2, 3, or 4. The data for each of these rules are stored in common block CINT by subroutine SETINT, which is given below.

This routine defines the values of the parameters required for the numerical integration of element matrices and vectors. These data are problem-independent and so require no reading of data records. As given below, SETINT defines data that allow numerical integration using Gaussian quadrature rules of order 1 through 4.

```
          SUBROUTINE SETINT
          COMMON/CINT/XI(4,4),W(4,4)
C.....GAUSSIAN QUADRATURE ORDER 1
          XI(1,1)=0.
          W(1,1)=2.
C.....GAUSSIAN QUADRATURE ORDER 2
          XI(1,2)=-1./SQRT(3.)
          XI(2,2)=-XI(1,2)
          W(1,2)=1.0
          W(2,2)=W(1,2)
C.....GAUSSIAN QUADRATURE ORDER 3
          XI(1,3)=-SQRT(3./5.)
          XI(2,3)=0.
          XI(3,3)=-XI(1,3)
          W(1,3)=5./9.
          W(2,3)=8./9.
          W(3,3)=W(1,3)
C.....GAUSSIAN QUADRATURE ORDER 4
          XI(1,4)=-0.8611363116
          XI(2,4)=-0.3399810436
          XI(3,4)=-XI(2,4)
          XI(4,4)=-XI(1,4)
          W(1,4)=0.3478548451
          W(2,4)=0.6521451549
          W(3,4)=W(2,4)
          W(4,4)=W(1,4)
          RETURN
          END
```

EXERCISE

Programming Assignment 3.1:

Write a FORTRAN program to perform numerical integration of the integrals given below. Your program should call subroutine SETINT to calculate the integration points XI and weights W, and should contain common block CINT. Use (3.3.6) to calculate the integrals. Calculate each integral using Gaussian quadrature of orders 1, 2, 3, and 4. Compare the numerical results with exact values.

$$\mathcal{I}_1 = \int_0^1 (x + x^2 + x^3)\, dx$$

$$\mathcal{I}_2 = \int_0^\pi \sin x\, dx$$

$$\mathcal{I}_3 = \int_0^5 \frac{1}{\sqrt{x}}\, dx$$

3.3.2 Shape-Function Routine—SHAPE

The values of the shape functions $\hat{\psi}_i(\xi)$ and their derivatives with respect to the element coordinate $d\hat{\psi}_i/d\xi$ are required in the evaluation of the integrals in (3.3.2) and (3.3.3). In CODE1, as in most finite element programs, these values are supplied by a shape function subroutine SHAPE, which is given below. This routine is called, in CODE1, by the element matrix calculation routine ELMAT and by routine EVAL in the postprocessor.

SHAPE calculates the values of the shape functions $\hat{\psi}_i$ and their derivatives $d\hat{\psi}_i/d\xi$ with respect to the master element coordinate at a specified value of ξ. This routine is supplied with the number of nodes (and therefore the number of shape functions) N and the value of ξ at which the values are to be calculated. The calculated values are returned in arrays PSI and DPSI.

```
       SUBROUTINE SHAPE(XI,N,PSI,DPSI)
       DIMENSION PSI(1),DPSI(1)
       IF(N.LT.2.OR.N.GT.4) GO TO 99
       GO TO (99,10,20,30) N
C..... LINEAR SHAPE FUNCTIONS
10     PSI(1)=.5*(1.-XI)
       PSI(2)=.5*(1.+XI)
       DPSI(1)=-.5
       DPSI(2)= .5
       RETURN
```

```
C..... QUADRATIC SHAPE FUNCTIONS
20        PSI(1)=XI*(XI-1.)*.5
          PSI(2)=1.-XI**2
          PSI(3)=XI*(XI+1.)*.5
          DPSI(1)=XI-.5
          DPSI(2)=-2.*XI
          DPSI(3)=XI+.5
          RETURN
C..... CUBIC SHAPE FUNCTIONS
30        PSI(1)=9./16.*(1./9.-XI**2)*(XI-1.)
          PSI(2)=27./16.*(1.-XI**2)*(1./3.-XI)
          PSI(3)=27./16.*(1.-XI**2)*(1./3.+XI)
          PSI(4)=-9./16.*(1./9.-XI**2)*(1.+XI)
          DPSI(1)=-9./16.*(3.*XI**2-2.*XI-1./9.)
          DPSI(2)=27./16.*(3.*XI**2-2./3.*XI-1.)
          DPSI(3)=27./16.*(-3.*XI**2-2./3.*XI+1.)
          DPSI(4)=-9./16.*(-3.*XI**2-2.*XI+1./9.)
          RETURN
99        WRITE(6,100)N
100       FORMAT(27H0 ERROR IN CALL TO SHAPE,N=I3)
          STOP
          END
```

EXERCISES

Programming Assignment 3.2:

1. Write a short FORTRAN program that calls subroutine SHAPE for several values of ξ such that $-1 \leq \xi \leq 1$ for linear, quadratic, and cubic shape functions. Calculate for each ξ and each $N = 2, 3, 4$

$$a = \sum_{i=1}^{N} \hat{\psi}_i(\xi) \qquad \text{and} \qquad b = \sum_{i=1}^{N} \frac{d\hat{\psi}_i(\xi)}{d\xi}$$

    and verify that $a = 1$ and $b = 0$. These tests are necessary (but not sufficient) to ensure correct calculation in SHAPE.

2. Write a short FORTRAN program utilizing subroutine SHAPE to calculate values of the finite element interpolant (see Section 2.6) of the function $g : g(x) = \sin \pi x$ for $0 \leq x \leq 1$. Use two elements with linear shape functions. Evaluate $g_h(x)$ and $g(x)$ at 10 equally spaced points in each element. Print the values of x, $g_h(x)$, $g(x)$, and $|g(x) - g_h(x)|$ at each point. Note that since the shape functions $\hat{\psi}_i$ are written as functions of ξ, where $-1 \leq \xi \leq 1$, the program must select the value of ξ at which calculations are to be made and then calculate the corresponding value of x from (3.3.5). Repeat this calculation for quadratic and cubic elements.

3. Using subroutine SETINT, extend the program for part **2** to calculate the mean-square norm of the error in the interpolations. Calculate and print the value

$$\| g - g_h \|_0 = \left\{ \int_0^1 [g(x) - g_h(x)]^2 \, dx \right\}^{1/2}$$

for linear, quadratic, and cubic elements. Use Gaussian integration of order 4.

3.3.3   Element Routine—ELEM

The evaluation of (3.3.2) using numerical integration of the form (3.3.6) is performed in subroutine ELEM described below. The integrands in (3.3.2) contain the values of the material properties $k(x)$, $c(x)$, and $b(x)$ and the load $f(x)$. These values are to be supplied by subroutine GETMAT, which is called by ELEM. A description (but no coding) of GETMAT follows the coding of ELEM.

In the shape function routine SHAPE, values of the derivatives of $\hat{\psi}_i$ with respect to the element coordinate ξ are calculated. Since the integrand of (3.3.2) contains the derivative with *respect to the global coordinate x*, we use the chain rule to calculate these quantities at the integration points. Thus,

$$\frac{d\psi_i}{dx} = \frac{d\hat{\psi}_i}{d\xi}\frac{d\xi}{dx} = \frac{d\hat{\psi}_i}{d\xi}\frac{2}{x_2 - x_1} \tag{3.3.7}$$

The explicit formulas for evaluating (3.3.2), according to (3.3.6) and the previous discussion, are

$$k_{ij}^e = \frac{dx}{d\xi} \sum_{l=1}^{N_l} w_l \left[k(x_l) \frac{d\hat{\psi}_i(\xi_l)}{d\xi} \frac{d\hat{\psi}_j(\xi_l)}{d\xi} \frac{d\xi}{dx} \frac{d\xi}{dx} \right.$$
$$\left. + c(x_l)\psi_i(\xi_l) \frac{d\hat{\psi}_j(\xi_l)}{d\xi} \frac{d\xi}{dx} + b(x_l)\hat{\psi}_i(\xi_l)\hat{\psi}_j(\xi_l) \right] \tag{3.3.8}$$

and

$$f_i^e = \frac{dx}{d\xi} \sum_{l=1}^{N_l} w_l \hat{\psi}_i(\xi_l) f(x_l) \tag{3.3.9}$$

In (3.3.8) and (3.3.9), $x_l = x(\xi_l)$ is the value of x calculated at ξ_l from (3.3.5).

Comparison of statements **11** and **12** in ELEM with (3.3.9) and (3.3.8) indicate how the calculations are made. Definitions of the quantities used in (3.3.8) and (3.3.9) and the FORTRAN variable names for each are given below.

X1,X2	= coordinates x_1 and x_2 of the ends of the element
N	= number of nodal points (and shape functions) in the element
NL	= order of the integration rule
XI(L)	= ξ_l, the location of integration point, l
X	= the value of the global coordinate at an integration point
DX	= $dx/d\xi = (x_2 - x_1)/2$; Note that $d\xi/dx = (dx/d\xi)^{-1}$
EK(I,J)	= k^e_{ij}, entry in element stiffness matrix
EF(I)	= f^e_i, entry in element load vector
MAT	= material number
W(L)	= w_l, the integration weights
PSI(I),DPSI(I)	= $\hat{\psi}_i(\xi_l)$, $d\hat{\psi}_i(\xi_l)/d\xi$, the values of shape functions and their derivatives at integration points
XK,XC,XB,XF	= $k(x_l)$, $c(x_l)$, $b(x_l)$, $f(x_l)$, values of the material properties at $x_l = x(\xi_l)$

The data that are passed to ELEM through the calling sequence include endpoint coordinates X1 and X2, number of nodes N, material number MAT, and the integration rule data NL, XI, and W. The routine returns to the calling program the calculated arrays EK and EF. Note that for one-dimensional arrays transmitted through the calling sequence, the exact dimension of the array need not be specified, and that for two-dimensional arrays thus transmitted, only the first dimensions need be specified exactly. Transmission of an array by means of the call statement passes to the called subprogram the beginning address of the array. The dimensioning of the array in the called subprogram serves only to allow the accurate calculation of the location of elements of the array relative to its beginning and not to reserve the necessary storage space for the array. Dimensioning in ELEM, as given below, allows the routine to accommodate finite elements of variable order easily. Since the arrays PSI and DPSI originate in ELEM, they are dimensioned here to accommodate the maximum order (i.e., four-node (cubic) elements).

The values of the coefficients $k(x)$, $c(x)$, $b(x)$, and $f(x)$ at the integration points are obtained from subroutine GETMAT, which is supplied with the x coordinate of the integration point X and the material number MAT. The values of the shape functions PSI and of their derivatives DPSI with respect to the element coordinate are returned by subroutine SHAPE, which is supplied with the integration point value of XI and the order of the element N.

```
      SUBROUTINE ELEM(X1,X2,N,EK,EF,NL,XI,W,MAT)
      DIMENSION EK(N,1),EF(1),XI(1),W(1)
      DIMENSION PSI(4),DPSI(4)
      DX=(X2-X1)/2.
C.....INITIALIZE ELEMENT ARRAYS
      DO 10 I=1,N
      EF(I)=0.0
      DO 10 J=1,N
10    EK(I,J)=0.
C.....BEGIN INTEGRATION POINT LOOP
      DO 20  L=1,NL
      X=X1+(1.+XI(L))*DX
      CALL GETMAT(MAT,X,XK,XC,XB,XF)
      CALL SHAPE(XI(L),N,PSI,DPSI)
      DO 20  I=1,N
11    EF(I)=EF(I)+PSI(I)*XF*W(L)*DX
      DO 20  J=1,N
12    EK(I,J)=EK(I,J)+(XK*DPSI(I)*DPSI(J)/DX/DX
     .            +XC*PSI(I)*DPSI(J)/DX+XB*PSI(I)*PSI(J))*W(L)*DX
20    CONTINUE
      RETURN
      END
```

3.3.4 Material-Property Routine—GETMAT

This routine calculates the values of the material properties (coefficients in the differential equation) at a given point in a given material. The value of X and the material number MAT are supplied in the calling statement, and the values of the properties XK, XC, XB, and XF are returned through the argument list. This routine operates in conjunction with routine RMAT, which reads and stores in common block CMATL the data used to calculate the properties. In the present case of constant coefficients, the coding is trivial. The call to GETMAT is given in the description of routine ELEM. GETMAT contains common block CMATL.

EXERCISE

Programming Assignment 3.3:
    Write a short FORTRAN program to call ELEM and print the element stiffness matrix and load vector. The program must call SETINT before the call to ELEM.

Routines SHAPE and GETMAT must also be included in the program. Use integration of order 3. Check several of the entries in $\mathbf{k}^e$ and $\mathbf{f}^e$ by hand calculation.

1. Linear element:

$$x_1 = \tfrac{1}{3},\ x_2 = \tfrac{2}{3},\ k(x) = -1,\ c(x) = 0,\ b(x) = 1,\ f(x) = x$$

(Note that this is the model problem of Chapter 1.)

2. Quadratic element:

$$x_1 = 0,\ x_2 = 2,\ k(x) = 2,\ c(x) = \tfrac{1}{2},\ b(x) = 0,\ f(x) = 1$$

Note that a different version of GETMAT must be used for each of the cases 1 and 2.

3.4 PREPROCESSING ROUTINES

The preprocessing unit of CODE1, as in any finite element code, exists to supply the data required for the calculation of the stiffness matrix and load vector for the problems to be solved. The definition of the various groups of data is made in subroutines called by routine PREP. For most of these routines, we leave the coding as exercises for the reader. As a general rule, each of the data-reading routines should perform the following operations:

1. Read the required data according to a convenient format.
2. Store the data in the appropriate common block.
3. Print the data in an easily interpreted format.

3.4.1 Preprocessor Calling Routine—PREP

This routine reads one data record which contains the title (up to 80 alphanumeric characters) of the problem. Since CODE1 is designed to process more than one problem, a new title is expected after the last data record in a problem. Normal termination of execution will occur when the last data record in a problem is followed by a record containing "END" in columns 1 through 3.

After reading and printing the problem title, PREP calls in turn each of the data-reading routines.

```
        SUBROUTINE PREP
        COMMON/CCON/TITLE(20)
        READ(5,100)TITLE
        IF(TITLE(1).EQ.3HEND) GO TO 99
        WRITE(6,101) TITLE
        CALL RCON
        CALL RNODE
        CALL RELEM
        CALL RMAT
        CALL RBC
        RETURN
99      STOP
100     FORMAT(20A4)
101     FORMAT(1H1,20A4)
        END
```

We describe next the structure and purpose of each of the subroutines called from subroutine PREP.

3.4.2   Control Data : Routine—RCON

Routine RCON reads the values of the control parameters NNODE, NELEM, NMAT, and NPOINT. (See Fig. 3.3 for definitions of these parameters.) Each value should be checked to see that it lies within the allowable range given below. If an error is detected, an error message should be printed and execution terminated. Finally, the values of all parameters should be printed using formats that clearly identify each value printed. The allowable range of control parameters follows:

$$2 \leq NNODE \leq 100$$
$$1 \leq NELEM \leq 99$$
$$1 \leq NMAT \leq 5$$
$$0 \leq NPOINT \leq 5$$

RCON contains common block CCON.

3.4.3   Nodal-Point Coordinate Definition : Routine—RNODE

Routine RNODE reads data that allow the definition of the x-coordinates of all nodal points in the problem. The nodes are numbered consecutively from 1 through NNODE (whose value is available in common block CCON).

The coordinates are stored in array X(100) and transmitted, as needed, through common block CNODE. There is a wide variation in the degree of sophistication possible in this, the grid-generation phase of finite element codes. In the simple example code given below, nodes are generated with equal spacing in various sections of the grid. A few checks are made on the input data to detect errors.

```
        SUBROUTINE RNODE
        COMMON/CCON/TITLE(20),NNODE
        COMMON/CNODE/X(100)
        REAL LARGE
C....   INITIALIZE COORDINATE ARRAY
        LARGE=1.E20
        DO 10 I=1,NNODE
10      X(I)=LARGE
        READ(5,100) NREC
C....   BEGIN LOOP TO READ NODAL POINT DATA
        DO 30  NR=1,NREC
        READ(5,100)N1,N2,X1,X2
        IF(N1.LT.1.OR.N2.GT.NNODE) GO TO 99
        IF(N2.NE.0) GO TO 11
        X(N1)=X1
        GO TO 30
11      DN=N2-N1
        DX=(X2-X1)/DN
        XX=X1-DX
C....   BEGIN LOOP TO DEFINE NODAL POINT COORDINATES
        DO 20  I=N1,N2
        XX=XX+DX
20      X(I)=XX
30      CONTINUE
C....   PRINT NODAL POINT COORDINATES
        WRITE (6,101) (I,X(I),I=1,NNODE)
        DO 40 I=1,NNODE
        IF(X(I).EQ.LARGE) GO TO 99
40      CONTINUE
        RETURN
99      WRITE(6,102)
        STOP
100     FORMAT(2I5,2F10.0)
101     FORMAT(9H0NODE NO.,I6,5X,12HX-COORDINATE,F11.3)
102     FORMAT(42H0-----ERROR IN NODAL POINT COORDINATE DATA)
        END
```

3.4.4 Element Data Definition: Routine—RELEM

Routine RELEM reads the data that allow the definition of all the elements in the finite element model. The elements are numbered consecutively from 1 through NELEM. The data that define element number I are:

KIND(I) = kind of element (degree of polynomial in the shape functions)

$$\text{KIND} = \begin{cases} 1 \text{ for linear elements} \\ 2 \text{ for quadratic elements} \\ 3 \text{ for cubic elements} \end{cases}$$

MAT(I) = number of the material that constitutes the element

NODES(J,I) J = 1, 2, . . . , (KIND(I) + 1) = nodal
point numbers of nodes in element.

These data are stored in the arrays in common block CELEM and transmitted as needed through this block. RELEM should read data records that allow the definition of the data described above. It is, of course, not desirable to require a separate data record to be read for each element being defined. The following format of data records will allow the complete specification of element data for reasonable user effort.

Data Record Format for Element Definitions

Record 1: NREC   = number of element data records
Record 2: NINTI, KINDI, MATI, NODEI, NUMBER

  NINTI   = order of the integration rule

  KINDI   = value of KIND for elements to be generated

  MATI   = value of MAT for elements to be generated

  NODEI   = nodal point number for first node of first element to be generated

  NUMBER = number of elements to be generated in sequence; element number and nodal point numbers are incremented as elements are generated; KIND, NINT, and MAT are constant for all the elements generated by this record

103

The coding of RELEM should cause record 1 to be read, then a loop entered within which record 2 is read and the appropriate number of elements generated. After the outer loop has been executed NREC times, the data for all the elements in the problem should be printed. RELEM contains common blocks CCON and CELEM.

3.4.5 Material-Property Definition: Routine—RMAT

This routine reads the data that allow the definition of the material properties $k(x)$, $c(x)$, $b(x)$, and $f(x)$. A variety of methods could be used to describe the variation of the functions—tables, interpolating polynomials, and so on. In CODE1 we consider materials within which each property is constant. The coding of RMAT in this case is straightforward. For each material, say material I for I = 1, 2, . . . , NMAT, the values of the properties are read, stored in the CMATL common block, and finally printed. RMAT contains common blocks CCON and CMATL.

3.4.6 Boundary-Condition Definition: Routine—RBC

This routine reads data that define the boundary condition at each end of the region of interest. For each boundary condition, the data consist of a flag KBC and values VBC. The meaning of these variables is shown below. Note that I = 1 for the left end and I = 2 for the right end.

$$\text{KBC(I)} = \begin{cases} 1\text{—essential (Dirichlet) boundary condition} \\ 2\text{—flux (Neumann) boundary condition} \\ 3\text{—natural boundary condition} \end{cases}$$

$$\text{VBC(1,I)} = \begin{cases} u_I & \text{for KBC(I)} = 1 \\ \sigma_I & \text{for KBC(I)} = 2 \\ p_I & \text{for KBC(I)} = 3 \end{cases}$$

$$\text{VBC(2,I)} = \begin{cases} 0 & \text{for KBC(I)} = 1 \text{ or } 2 \\ u_I & \text{for KBC(I)} = 3 \end{cases}$$

Routine RBC also reads the data to define point loads (if any). For each of the NPOINT loads, the nodal point number at which the load acts, KPOINT, and the value of the load, POINT, are to be read.

The data read by RBC are stored in the arrays in CBC and transmitted through this block. The boundary condition and point load data should, of course, be printed after they have been read. **RBC** contains common blocks **CCON** and **CBC**.

EXERCISE

Programming Assignment 3.4:
Complete the coding of the preprocessing routines. Compile and correct FORTRAN errors. Prepare a user's guide containing a description of each data record required as input. Prepare a data set for the model problem of Chapter 1. Run the data set through the preprocessor and check the output to verify the correct operation of the preprocessor.

3.5 FINITE ELEMENT CALCULATION ROUTINES

This unit consists of the calling routine **PROC** and the routines that calculate, assemble, modify for boundary conditions, and solve the finite element equations $\mathbf{Ku} = \mathbf{F}$.

3.5.1 Processor Calling: Routine—PROC

This routine contains the calls to the various components of the analysis procedure, which, in **CODE1**, consists simply of successive calls to routines **FORMKF**, **APLYBC**, and **SOLVE**.

3.5.2 Formation of K and F: Routine—FORMKF

Routine **FORMKF** directs the calculation of element stiffness matrices $\mathbf{k}^e$ and load vectors $\mathbf{f}^e$ by routine **ELEM** and their accumulation into the arrays **K** and **F** by routine **ASSMB**. It is in **FORMKF** that the element-by-element nature of finite element calculations is most clearly seen to be computationally advantageous. This routine begins by zeroing the global stiffness matrix, **GK**, and load vector **GF**. Then, element by element, the routines that calculate and assemble the element matrices, **EK**, and vectors, **EF**, are called.

The element routine **ELEM** (see Section 3.3.3) requires the coordinates of the end nodes (the first and last) of the element; the location and size of the element arrays **EK** and **EF**; the integration data **NL**, **XI**, and **W**; and the material number of the element. This can be coded as follows:

```
DO 20 NEL=1,NELEM
N=KIND(NEL)+1
I1=NODES(1,NEL)
I2=NODES(N,NEL)
I3=NINT(NEL)
CALL ELEM(X(I1),X(I2),N,EK,EF,I3,XI(1,I3),W(1,I3),MAT(NEL))
    .
    .
    .
20      CONTINUE
```

The assembly routine ASSMB (see Section 3.5.3) requires the element and global matrices and the nodal point numbers (equation numbers) for the nodes in the element. Since the node numbers for each element are stored consecutively in the array NODES (FORTRAN stores two-dimensional arrays by columns), it suffices to transmit through the calling sequence only the location of the first node number of the element. In order to allow the assembly routine to remain independent of the size of the various arrays it uses, these arrays are transmitted through the calling sequence together with their dimensions. The assembly routine may be called by the following sequence of code.

```
NG=100
DO 20 NEL=1,NELEM
    .
    .
    .
        CALL ASSMB(EK,EF,N,NODES(1,NEL),GK,GF,NG)
20      CONTINUE
```

The completion of FORMKF is left as an exercise. FORMKF contains common blocks CCON, CNODE, CELEM, CINT, and CMATRX.

3.5.3 Assembly Routine—ASSMB

In ASSMB, the contributions to the global matrix **K** and load vector **F** from a single element are added to the accumulated contributions from other elements. Before ASSMB is called the first time, the global arrays are set to zero. After ASSMB has been called for each element, the global arrays are complete.

In CODE1 we are solving a scalar differential equation. There is one nodal point value for each node and one equation. In an element there are N nodal values and N equations; the element matrix k^e is N × N. The list of N nodal point numbers, supplied to ASSMB through the argument list in local array NODE, specifies the rows and columns of **K** into which the entries of k^e are to be accumulated. The coding of ASSMB given below is a concise way of defining the assembly procedure common to all finite element codes.

```
        SUBROUTINE ASSMB(EK,EF,N,NODE,GK,GF,NG)
        DIMENSION EK(N,1),EF(1),NODE(1),GK(NG,1),GF(1)
        DO 10 I=1,N
        IG=NODE(I)
C..... ASSEMBLE GLOBAL VECTOR GF
        GF(IG)=GF(IG)+EF(I)
        DO 10 J=1,N
        JG=NODE(J)
C..... ASSEMBLE GLOBAL STIFFNESS MATRIX GK
10      GK(IG,JG)=GK(IG,JG)+EK(I,J)
        RETURN
        END
```

EXERCISES

3.5.1 Construct the array NODE as it would be required by subroutine ASSMB for each element in the following finite element model:

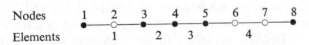

3.5.2 Place an x in the appropriate entry of the 8 × 8 stiffness matrix **K** for every number that would be added to **K** by routine ASSMB. Use the NODE arrays constructed in Exercise 1.

3.5.4 Boundary Conditions: Routine—APLYBC

The routine APLYBC makes the modifications to the global matrix **K** and vector **F** required to impose the boundary conditions at each end of the domain. The treatment of point loads is also made in this routine. The modifications for boundary conditions of the three kinds allowed are described below. APLYBC contains common block CCON, CBC, and CMATRX.

1. **Essential Boundary Condition** (KBC = 1): In this case the value of u at the first (or last) node is to have the value u_a (or u_b) given by VBC(1,1) (or VBC(1,2)) (recall (3.2.2)). To impose this condition we multiply the first (last) column of GK by u_a (u_b) and then subtract the result from GF. Then the first (last) row and column of GK are set to zero. Finally, the value 1 is placed in the first (last) diagonal of GK and the first (last) component of GF set equal to u_a (u_b). This compli-cated procedure is rarely done in engineering practice but is included here because it corresponds exactly to the variational statement of the problem. That is, we use u_a (u_b) for the known value of u in all equa-tions (transferring known quantities to the right-hand side of the equation); we multiply the first (last) equation by the *known* value (zero) of the test function at the node; and finally we use the vacant position in GK and GF to make sure that in the list of nodal point values $u = u_a$ (u_b). This treatment of the essential boundary condi-tions is implemented in a separate routine DRCHLT: first, because of its complexity, and second, because it will be replaced by an efficient alternative. (See Section 3.8.1, p. 121.)

2. **Flux Boundary Condition** (KBC = 2): In this case the value of the flux σ_a (or σ_b) is given by VBC(1,1) (or VBC(1,2)). According to the variational statement of the boundary-value problem, this is a known value multiplied by the unknown value of the test function at the first (last) node. Thus, we simply add σ_a (or σ_b) to the first (last) component of GF.

3. **Natural Boundary Condition** (KBC = 3): In this case the boundary coefficient p_a (or p_b) is given by VBC(1,1) (or VBC(1,2)) and the "environmental value" of u, namely u_a (or u_b), is given by VBC(2,1) (or VBC(2,2)). According to the variational statement, p_a multiplies the unknown nodal u and the value of the test function at the first (last) node. Thus, we add the value p_a (p_b) to the first (last) diagonal of GK. The quantity $p_a u_a$ ($p_b u_b$) multiplies the value of the test func-tion at the node, so this quantity is added to the GF vector.

3.5.5 *Essential Boundary Condition: Routine—DRCHLT*

```
      SUBROUTINE DRCHLT(VAL,N,NNODE)
      COMMON/CMATRX/GK(100,100),GF(100)
      DO 10 I=1,NNODE
C.....MODIFY RIGHT HAND SIDE VECTOR (VAL = VBC)
      GF(I)=GF(I)-GK(I,N)*VAL
```

```
C.....MODIFY ROW N AND COLUMN N OF STIFFNESS MATRIX
      GK(I,N)=0.
10    GK(N,I)=0.
      GK(N,N)=1.
      GF(N)=VAL
      RETURN
      END
```

3.5.6 Equation Solving: Routine—SOLVE

This routine calls the routines that calculate the solution to the finite element equations $Ku = F$. The solution is achieved by performing Gaussian elimination to reduce the K matrix (array GK) to upper-triangular form in routine TRI. Then the forward elimination steps are applied to the right-hand-side vector F (array GF) and the solution vector u (array U) is calculated by back-substitution in routine RHSUB. These two routines could easily be combined into a single routine. For use in CODE1, this single routine would function as well as the two separate ones. In many applications of finite elements, however, it is advantageous to separate these parts of the calculation so that the most expensive part, the triangularization of K, can be done once and several successive right-hand sides processed subsequently. Applications in which this is desirable include time-dependent and/or nonlinear problems.

```
SUBROUTINE SOLVE
COMMON/CCON/TITLE(20),NNODE
COMMON/CMATRX/GK(100,100),GF(100)
COMMON/CNODE/X(100),U(100)
CALL TRI(GK,100,NNODE)
CALL RHSUB(GK,U,GF,100,NNODE)
RETURN
END
```

3.5.7 Reduction of K to Triangular Form: Routine—TRI

This routine performs standard Gaussian elimination (row operations) on the matrix K. No pivoting (searching for large entries to divide by) is done. Although, with the storage of the full matrix, this could be easily accomplished, it is not required in the solution of most finite element equations. Furthermore, in the more efficient forms of matrix storage used in typical application codes, pivoting is complicated and seldom used. The algorithms coded below and in routine RHSUB are the simplest possible and are not typical of those used in practice. (See Section 3.8.1, p. 118, for a more typical form of matrix storage.)

```
        SUBROUTINE TRI(A,N,M)
        DIMENSION A(N,N)
        M1=M-1
        TINY=1.E-30
C . . . . . ELIMINATE DEGREE OF FREEDOM I
        DO 30   I=1,M1
C . . . . . CHECK FOR EXCESSIVELY SMALL PIVOT
        IF(ABS(A(I,I)).LT.TINY) GO TO 99
        J1=I+1
C . . . . . MODIFY ROW J
        DO 20   J=J1,M
        IF(A(J,I).EQ.0.) GO TO 20
        FAC=A(J,I)/A(I,I)
        DO 10 K=J1,M
10      A(J,K)=A(J,K)-A(I,K)*FAC
20      CONTINUE
30      CONTINUE
        RETURN
99      WRITE (6,100)I,A(I,I)
100     FORMAT(36H0REDUCTION FAILED DUE TO SMALL PIVOT/
        .19H EQUATION NO.,PIVOT,I5,E10.3)
        STOP
        END
```

3.5.8 Forward Elimination of F and Calculation of u: Routine—RHSUB

Routine RHSUB does the forward substitution on the right-hand-side vector, array B, repeating the row operations that were done in TRI, and then performs the back substitution to calculate the values of **u**, array X. We note that the numbers in the lower-triangular part of the A array still contain the information necessary to perform these operations. It is also notable that vector B is destroyed by this routine and that the solution, array X could be calculated in the B array, thus saving storage locations. This could be done by calling RHSUB with the second and third arguments equal. For example, CALL RHSUB(GK,GF,GF,100,NNODE) would produce the solution in the GF array. This feature is common to elimination methods and is frequently utilized in applications where storage is at a premium.

```
      SUBROUTINE  RHSUB(A,X,B,N,M)
      DIMENSION  A(N,N),X(1),B(1)
      M1=M−1
C.....BEGIN FORWARD REDUCTION OF RIGHT HAND SIDE
      DO 20  I=1,M1
      J1=I+1
      DO 10 J=J1,M
10    B(J)=B(J)−B(I)*A(J,I)/A(I,I)
20    CONTINUE
C.....BEGIN BACK SUBSTITUTION
      X(M)=B(M)/A(M,M)
      DO 40 I=1,M1
      IB=M−I
      J1−IB+1
      DO 30  J=J1,M
30    B(IB)=B(IB)−A(IB,J)*X(J)
      X(IB)=B(IB)/A(IB,IB)
40    CONTINUE
      RETURN
      END
```

EXERCISES

Programming Assignment 3.5:

1. Complete the coding for, compile, and correct FORTRAN errors in the
    processor routines of Section 3.5 (including element calculation routines of
    Section 3.3). Compile these together with main program and preprocessor
    routines. Add temporary print statements in FORMKF to print the active
    portion of the stiffness matrix GK and load vector GF after each call to
    ASSMB and in PROC after the call to APLYBC (the active portion of GK is
    GK(I,J), $1 \leq I, J \leq$ NNODE).

2. Prepare a data set for the model problem of Chapter 1. Run this data set
    and verify that the correct results have been obtained after each step in the
    assembly and modification of **K** and **F**, and that the correct nodal point
    values **u** are calculated in SOLVE. Print array U in SOLVE. After verification,
    deactivate the temporary print statements.

3.6 POSTPROCESSING

The variety of features found in postprocessing routines in finite element
codes is extensive. In CODE1, we include the evaluation of exact solutions
and error norms. Our inclusion of these features makes CODE1 a useful

tool for numerical experiments to study errors in finite element solutions (see Section 3.8.2).

3.6.1 Postprocessing Control: Routine—POST

```
        SUBROUTINE POST
        COMMON/CCON/TITLE(20),NNODE,NELEM
        DIMENSION OUTPTS(10)
1       READ (5,100)CMMND,NEL1,NEL2
        IF(CMMND.EQ.3HOUT) GO TO 10
        IF(CMMND.EQ.3HERR) GO TO 20
        GO TO 99
C....   OUTPUT SOLUTION AT SELECTED POINTS
10      READ(5,101)NPTS,(OUTPTS(I),I=1,NPTS)
        WRITE(6,102) NPTS,(OUTPTS(I),I=1,NPTS)
102     FORMAT(X,I3,15H OUTPUT POINTS:,10F6.3)
        IF(NEL1+NEL2.GT.0) GO TO 11
        NEL1=1
        NEL2=NELEM
11      WRITE(6,103)TITLE
        DO 13 NEL=NEL1,NEL2
        WRITE(6,104) NEL
        DO 13 NPT=1,NPTS
        CALL EVAL(NEL,OUTPTS(NPT),XX,UH,UX,SIGH,SIGX)
        DU=UX-UH
        DSIG=SIGX-SIGH
        WRITE(6,105) XX,UH,UX,DU,SIGH,SIGX,DSIG
13      CONTINUE
        GO TO 1
C.....  CALL ERROR NORM ROUTINE (IF EXACT SOLUTION IS
C.....  AVAILABLE)
20      CALL ENORMS
        GO TO 1
99      RETURN
100     FORMAT(A3,2I5)
101     FORMAT(I5,10F5.4)
103     FORMAT(1H1,20A4/
       .4X,1HX,8X,1HU,10X,1HU,10X,1HU,7X,7H-KDU/DX,4X,
       .7H-KDU/DX,4X,7H-KDU/DX/
       .12X,3HF E,7X,5HEXACT,6X,5HERROR,7X,3HF E,7X,5HEXACT,
       .6X,5HERROR)
104     FORMAT(12H0ELEMENT,NO.,I3)
105     FORMAT(F8.4,6E11.3)
        END
```

Data Records Required by Routine **POST**

Record 1: CMMND,NEL1,NEL2

CMMND = alphanumeric command that directs the routine to evaluate solution, calculate error norms, or return to beginning of code. Value of CMMND =
OUT—evaluate and print solution
ERR—evaluate and print error norms
END—return to beginning of code

NEL1,NEL2 = element numbers between which solution is to be printed (inclusive); default values are NEL1 = 1, NEL2 = NELEM; NEL1 and NEL2 are not required by ERR

Following a Record 1 with command = OUT, there must appear a Record 2.

Record 2: NPTS,(OUTPTS(I),I=1,NPTS)

NPTS = number of points in each element at which solution is to be printed—maximum of 10

OUTPTS = values of the element coordinate ξ at which solution is to be printed

3.6.2 Evaluation of Finite Element and Exact Solutions: Routine—EVAL

The routine EVAL, called by POST and ENORMS, calculates the value of the finite element solution UH and the stress (flux) SIGH at a given point in a given element. The element number NEL and element coordinate XI at which the solution is to be calculated are transmitted to EVAL through the calling sequence. Then routine EXACT is called to obtain exact values of the solution UX and stress SIGX.

```
        SUBROUTINE EVAL(NEL,XI,XX,UH,UX,SIGH,SIGX)
        COMMON/CNODE/X(100),U(100)
        COMMON/CELEM/KIND(99),MAT(99),NODES(4,99),NINT(99)
        DIMENSION PSI(4),DPSI(4)
        N=KIND(NEL)+1
        CALL SHAPE(XI,N,PSI,DPSI)
C.....CALCULATE UH AND DUH/DX FROM SHAPE FUNCTIONS
        UH=0.
        DUHDX=0.
```

```
          DO 10 I=1,N
          I1=NODES(I,NEL)
          IF(I.EQ.1) X1=X(I1)
          IF(I.EQ.N) X2=X(I1)
          UH=UH+PSI(I)*U(I1)
10        DUHDX=DUHDX+DPSI(I)*U(I1)
          DX=(X2-X1)*.5
          XX=X1+(1.+XI)*DX
          CALL GETMAT(MAT(NEL),XX,XK,XC,XB,XF)
          SIGH=-DUHDX*XK/DX
C.....CALCULATE EXACT VALUES UX AND SIGX
          CALL EXACT(XX,UX,SIGX)
          RETURN
          END
```

3.6.3 Error-Norm Calculation: Routine—ENORMS

This routine calculates and prints two measures of the error in the finite element solution. These are, first, the mean-square norm defined in equation (1.8.3), and the energy norm, which has the form

$$\|u - u_h\|_E = \left\{ \int_a^b [k(u' - u_h')^2 + b(u - u_h)^2] \, dx \right\}^{1/2} \qquad (3.6.1)$$

The integrals in the definitions of these norms are evaluated numerically, element by element.

In order to be sure that rapidly varying errors do not escape detection by the numerical integration, a large number of integration points are used in each element, together with a very simple quadrature rule (trapezoidal rule). At each integration point, the value of the finite element and exact solution and stress are calculated by EVAL.

```
          SUBROUTINE ENORMS
          COMMON/CCON/TITLE(20),NNODE,NELEM
          COMMON/CELEM/KIND(99),MAT(99),NODES(4,99)
          SQNORM=0.
          ENORM=0.
          DXI=2./49.
C.....LOOP ON ELEMENTS
          DO 50 I=1,NELEM
          XI=1.
          CALL EVAL(I,XI,X,UH,UX,SIGH,SIGX)
          X2=X
          XI=-1
C.....LOOP ON INTEGRATION POINTS
          DO 40 K=1,50
C.....GET EXACT AND APPROXIMATE VALUES OF U AND SIG
          CALL EVAL(I,XI,X,UH,UX,SIGH,SIGX)
          CALL GETMAT (MAT(I)X,XK,XC,XB,XF)
          W=1.
```

```
          XI=XI+DXI
          IF(K.NE.1) GO TO 10
          W=0.5
          X1=X
          GO TO 25
   10     IF(K.NE.50) GO TO 25
          W=0.5
          X2=X
   25     CONTINUE
          SQNORM=SQNORM+(UX-UH)**2*W
          ENORM=ENORM+(SIGX-SIGH)**2*W/XK+(UX-UH)**2*W*XB
          SQNORM=SQNORM*(X2-X1)/49.
   40     ENORM=ENORM*(X2-X1)/49.
          SQNORM=SQRT(SQNORM)
          ENORM=SQRT(ENORM)
          WRITE(6,100) TITLE,SQNORM,ENORM
  100     FORMAT(1H1,20A4,/,24H0 THE MEAN SQUARE ERROR   ,
         .43HIN THE FINITE ELEMENT SOLUTION   //UX-UH//0= ,E11.3,/,
         .50H0 THE ENERGY ERROR IN THE FINITE ELEMENT SOLUTION ,
         .18H          //UX-UH//E-,E11.3)
          RETURN
          END
   50     CONTINUE
```

3.6.4 Calculation of Exact Solution : Routine—EXACT

This routine must be modified by the user for each different boundary-value problem. For a given value of X transmitted to EXACT through the calling sequence, the exact values of the solution U and the stress SIG (the stress is given by $\sigma = -ku'$) are calculated. The coding shown below is an example of EXACT for the model problem of Chapter 1.

```
          SUBROUTINE EXACT(X,U,SIG)
   C..... MODEL PROBLEM
   C..... -D2U/DX2+U=X
   C..... U(X)=X-SINH(X)/SINH(1)
   C..... SIG(X)=-1+COSH(X)/SINH(1)
   C.....
          E=EXP(1.)
          EX=EXP(X)
          S1=E-1./E
          SX=EX-1./EX
          CX=EX+1./EX
          U=X-SX/S1
          SIG=-1.+CX/S1
          RETURN
          END
```

EXERCISES

Programming Assignment 3.6:
    Compile the postprocessor routines together with the remainder of CODE1. Run the data set for the model problem from Programming Assignment 3.5. Calculate the solutions (finite element and exact) at the nodal points and midpoints of each element (i.e., use $\xi = -1, 0, 1$ as output points). Calculate the error norms and verify that the values given in Chapter 1 are obtained.

Programming Assignment 3.7:
    Complete the user's guide that was begun in Programming Assignment 3.4 to include directions for use of the postprocessor. Give, as an appendix, the description of the finite element modeling of the model problem, including input data and printed output.

3.7 REMARKS ON THE DEVELOPMENT OF CODE1

The finite element computer program described in this chapter is, in some ways, typical of codes widely used for applications in many areas of engineering analysis. The organization of the calculations and the data, the ability of the code to treat any boundary-value problem in a fairly general class of problems, the ability provided the user to model the problem with elements of varying size and order, the calculation of element stiffness and load matrices by numerical integration, and the solution of the equations by elimination are, indeed, typical of application codes.

In some respects, however, CODE1 differs from the usual general-purpose code. Most obviously, CODE1 treats one-dimensional problems while the true potential of the finite element method is best realized in higher dimensions. The coding described for CODE1 fails to take into account the sparseness of the global stiffness matrix, leading to an extravagant waste of computer storage and computer time that could not be tolerated in application codes—especially in problems of higher dimension. Techniques that take full advantage of the special character of the sparse stiffness matrices in finite element problems have been highly developed. A detailed treatment of these is given in Volume III of this series. A simple modification of CODE1 that provides reasonably efficient storage and computation of the global stiffness matrix is treated in Section 3.8.1, p. 118.

In order to provide the reader with a tool for studying the effects of various finite element modeling as treated in the exercises in Section 3.8.2, we have included two features in CODE1 not usually found in typical application codes. The first is the allowance for quadrature rules of different order for different elements. The second is the inclusion of the calculation of the exact solution and the errors in the finite element solution.

3.8 EXTENSION AND APPLICATION OF CODE1

The exercises in this section are of three kinds: further development of CODE1, numerical experiments for studies of the finite element method, and modeling of physical problems.

3.8.1 Topics in Code Development

Verification: An important phase in the development of a general-purpose finite element program is the verification of the code. Verification consists of exercising the code on a set of problems that are within the design capabilities of the code, which test each of the various capabilities and whose solution is known (and as simple as possible).

Consider the construction of a set of problems that verify the capability of CODE1 to calculate correctly approximate solutions using all types of elements and boundary conditions in the code.

For example, to test quadratic elements, we can devise boundary-value problems whose solutions are a second-degree polynomial. In particular, let $u(x) = 1 + 2x - 3x^2$ and note that u is the solution of $-u'' = 6x$ that satisfies the following conditions at $x = 0$ and at $x = 1$.

$$
\begin{array}{ll}
x = 0 & x = 1 \\
u(0) = 1 & u(1) = 0 \\
\sigma(0) = -u'(0) = -2 & \sigma(1) = -u'(1) = 4 \\
\sigma(0) = 1[u(0) - 3] & \sigma(1) = 2[u(1) + 2]
\end{array}
$$

From these results, we can construct a variety of verification problems.

One such problem would be stated: find u so that

$$-u'' = 6x, \qquad 0 < x < 1$$

and

$$u(0) = 1, \qquad u'(1) = -4$$

It should be clear that the complete verification of even a simple program such as CODE1 is an arduous task. The prudent user of finite element codes will always devise and test a few verification problems when beginning to use a newly-developed or newly-acquired code.

EXERCISES

To run verification problems for CODE1 it is necessary to make appropriate changes in GETMAT and EXACT.

3.8.1 Make the necessary code changes, prepare a data set and run the quadratic verification problem. (Use at least second-order integration.)

3.8.2 Devise a verification problem to test cubic elements and natural boundary conditions. Run the problem.

Banded Matrix Storage. As noted in the discussion in Section 3.7, storing the full stiffness matrix **K** is neither necessary nor efficient. In this exercise, a modified scheme for the storage and manipulation of **K** (array GK in the code) is developed. Modification of CODE1 to include this scheme will induce a significant reduction in the storage requirements of the code. The storage and solution presented here exploit the *banded* character of the stiffness matrix.

Figure 3.4a shows schematically the location of the nonzero entries in

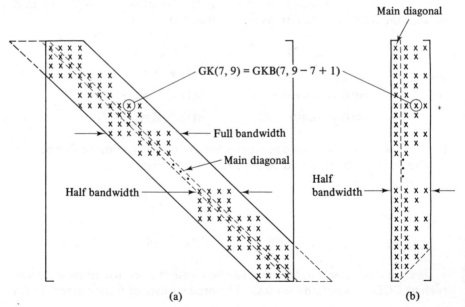

(a)   (b)

FIGURE 3.4 *Matrix storage.*

the array GK for a typical finite element model in CODE1 using cubic elements. Clearly, all the terms in **K** that are either nonzero initially or into which nonzero values are introduced by the elimination procedure lie within the *band* enclosed by solid lines in Fig. 3.4a. The width of this band is determined by the finite element model. In one-dimensional problems, the number of nodes in an element is the extent of the band on either side of the main diagonal, shown by dashed lines in Fig. 3.4a. For cubic elements the *bandwidth* is seven.

In general finite element models, the bandwidth can be determined as the maximum value over all the nodes of the nodal separations in a problem. For a given node, the nodal separation is the difference (plus one) between the largest node number in an element connected to the given node and the smallest node number so connected. For problems in two or three dimensions, the bandwidth may be considerable (sometimes a few hundred), whereas for one-dimensional problems it is always small. The scheme presented in this exercise, which takes advantage of the concentration of nonzero entries of **K** around the main diagonal, utilizes what are called banded storage and solution techniques.

Figure 3.4b shows the upper triangular part of **K** stored in banded form. In banded form, the array is two dimensional, as it is in full form, and there is a one-to-one relation between the rows in the two forms. The columns of the full matrix, however, have been skewed (i.e., pushed to the left), so that the main diagonal element in each row appears in the first column in the banded form. It is convenient to visualize the operations performed on the global stiffness matrix **K** as being done with the full storage form. However, in coding these operations, the addresses of the elements in the banded form will be used. These operations include the assembly of element stiffnesses $\mathbf{k}^e$, the modification for boundary conditions, and the elimination and substitution procedures. The correspondence of addresses between the two forms is obtained by inspection of Fig. 3.4. An entry that would be stored in row I and column J of the GK array is stored in row I and column $J - I + 1$ of the GKB array. This algorithm for calculating equivalent addresses makes the treatment of banded storage simple to code.

Here we treat modifications of CODE1 that will provide for treatment of problems with banded symmetric stiffness matrices only (i.e., problems for which c is identically zero). There is no conceptual difficulty in providing a similar treatment for nonsymmetric problems. We leave this as an additional exercise for the reader.

The modification of CODE1 for the inclusion of the banded form of matrix storage consists of the following steps:

1. Common block CMATRX should be changed in all routines in which it appears to read **COMMON/CMATRX/GKB(100,4),GF(100)**.

The initialization of these arrays in FORMKF must, therefore, be made compatible.

2. The assembly of element stiffness matrices in ASSMB should be changed so that only numbers in the upper-triangular part of **K** are stored in GKB, and, of course, the subscript JG of terms in GKB should be changed to agree with the preceding algorithm. These changes can be effected by making the inner DO loop in ASSMB read as follows:

```
      DO 10 J=1,N
      JG=NODE(J)−IG+1
      IF(JG.LE.0)   GO TO 10
      GKB(IG,JG)=GKB(IG,JG)+EK(I,J)
   10 CONTINUE
```

3. The location of the main diagonal elements of matrix GKB is always in column 1. References to the main diagonal in APLYBC and DRCHLT must be changed accordingly.

4. Routines TRI and RHSUB must be changed to take into account the banded storage. Routines TRIB and RHSB, given below, have these changes incorporated.

```
      SUBROUTINE TRIB(A,N,IB,L1,L2)
      DIMENSION A(L1,L2)
C.....THIS SUBROUTINE TRIANGULARIZES A BANDED AND SYMMETRIC
C.....MATRIX A
C.....ONLY THE UPPER HALF-BAND OF THE MATRIX IS STORED
C.....STORAGE IS IN THE FORM OF A RECTANGULAR ARRAY L1 × L2
C.....THE HALF-BAND WIDTH IS IB
C.....THE NUMBER OF EQUATIONS IS N
      DO 120 I=2,N
      M1=MIN0(IB−1,N−I+1)
      DO 120 J=1,M1
      SUM=0.0
      K1=MIN0(I−1,IB−J)
      DO 100 K=1,K1
  100 SUM=SUM+A(I−K,K+1)*A(I−K,J+K)/A(I−K,1)
  120 A(I,J)=A(I,J)−SUM
      RETURN
      END

      SUBROUTINE RHSB(A,X,B,N,IB,L1,L2)
      DIMENSION A(L1,L2),X(1),B(1)
C.....FOR THE LINEAR SYSTEM A*X=B WITH THE MATRIX A
C.....TRIANGULARIZED BY ROUTINE TRIB
```

```
C.....THIS ROUTINE PERFORMS THE FORWARD SUBSTITUTION
C.....INTO B AND BACK SUBSTITUTION INTO X
C.....THE HALF-BAND WIDTH OF A IS IB
C.....THE NUMBER OF EQUATIONS IS N
      NP1=N+1
      DO 110 I=2,N
      SUM=0.
      K1=MINO(IB-1,I-1)
      DO 100 K=1,K1
100   SUM=SUM+A(I-K,K+1)/A(I-K,1)*B(I-K)
110   B(I)=B(I)-SUM
C.....BEGIN BACK-SUBSTITUTION
      X(N)=B(N)/A(N,1)
      DO 130 K=2,N
      I=NP1-K
      J1=I+1
      J2=MINO(N,I+IB-1)
      SUM=0.0
      DO 120 J=J1,J2
      MM=J-J1+2
120   SUM=SUM+X(J)*A(I,MM)
130   X(I)=(B(I)-SUM)/A(I,1)
      RETURN
      END
```

EXERCISE

3.8.3　Make the necessary modifications to CODE1 to incorporate banded matrix storage. Verify the accuracy of the modified code by rerunning the data set from Exercise 3.8.1.

Essential Boundary Conditions: The modification of the assembled stiffness matrix **K** and right-hand side vector **F** to account for the essential boundary conditions can be coded more simply than by the method described in Section 3.5.4 and implemented in subroutine DRCHLT of Section 3.5.5. In this approach the simpler method of imposing the essential boundary conditions is to be incorporated into CODE1.

The method underlying this, which is often referred to as a *penalty* method, is based on the consideration of an essential boundary condition as a limiting case of a natural boundary condition. Suppose that the essential boundary condition to be imposed is

$$u(0) = u_0 \qquad\qquad (3.8.1)$$

If, in the general form of the natural boundary condition given by (2.2.2), we set $\gamma_0 = \beta_0 u_0$ then the form of the natural boundary condition becomes

$$\alpha_0 \frac{du(0)}{dx} + \beta_0 u(0) = \beta_0 u_0 \tag{3.8.2}$$

By choosing β_0 to be a number that is large compared to α_0, (3.8.2) approximates (3.8.1). If β_0 is also large compared to the entries of **K**, then adding (3.8.2) to the first equation of the assembled linear system, as in (2.7.11), we can assure that (3.8.2) will be satisfied. The original first equation is no longer satisfied—in fact, if β_0 is large enough, the original first equation is completely dominated by the addition of (3.8.2). In the jargon of optimization theory, the first equation is said to be penalized by the condition (3.8.2) and the number β_0 is referred to as a *penalty number*. Recalling that our original method of treating the essential boundary condition is to replace the first equation by (3.8.1), we see that the penalty method achieves our objective as long as β_0 is sufficiently large.

The penalty method is not only easy to implement and mathematically sound but it is also easy to motivate by physical arguments. In case the boundary-value problem arises from, say, the loading of an elastic string, the natural boundary condition (2.2.2) represents the effect of an elastic support with stiffness β_0. The essential boundary condition (3.8.1) corresponds to a rigid support. By choosing a large enough value of support stiffness we can simulate, as accurately as we want, the rigid support condition. In actual computations we need only choose the penalty number β_0 a few orders of magnitude larger than other terms in **K**. It is important to note that the results are insensitive to the actual choice of β_0. In the coding given below for subroutine DRCHLT a value of β_0 is chosen to be 10^{10} times as large as the diagonal term of **K** to which β_0 is added.

It is worth noting that constraints other than essential boundary conditions can be imposed by the use of penalty methods. Any constraint which must be imposed by restricting the class of admissible functions, rather than being a consequence of the variational form of the boundary-value problem, is likely to lend itself to treatment by a penalty method. Examples of such constraints include the imposition of continuity conditions, the forcing of two or more nodal values of u_h to have the same value, and forcing the deformation of two- or three-dimensional fluid or solid bodies to occur without a volume change.

The simplicity of the implementation of the penalty method is illustrated in the coding, given below, for subroutine DRCHLT (c.f. Section 3.5.5):

```
SUBROUTINE DRCHLT (VAL,N,NNODE)
COMMON/CMATRX/GK(100,100),GF(100)
DATA PEN/1.E10/
BETAO=PEN*GK(N,N)
GK(N,N)=GK(N,N)+BETAO
GF(N)=GF(N)+BETAO*VAL
RETURN
END
```

EXERCISES

3.8.4   Work the model problem of Chapter 1 using the penalty method to enforce both of the boundary conditions. Write the assembled equations for all five degrees of freedom. Use $\beta_0 = 1000$ and compare the results with those given in Chapter 1. Observe how the penalty dominates the remainder of the terms in equations 1 and 5.

3.8.5   Use CODE1 with the modified version of DRCHLT given in this section to solve the model problem of Chapter 1. Vary the value of PEN from, say, 10^2 to 10^{10} and observe the results. Draw conclusions.

3.8.2   *Numerical Experiments*

The function u defined below can be made to exhibit behavior which ranges from very smooth to almost discontinuous near $x = x_0$ by varying the values of the parameters x_0 and α.

$$u(x) = (1 - x)[\tan^{-1} \alpha(x - x_0) + \tan^{-1} (\alpha x_0)]$$

Note that

$$u'(x) = -[\tan^{-1} \alpha(x - x_0)] + [\tan^{-1} (\alpha x_0)] + \frac{(1 - x)\alpha}{1 + \alpha^2(x - x_0)^2}$$

and that u is a solution to the differential equation

$$-(k(x)u'(x))' = f(x)$$

where

$$k(x) = \frac{1}{\alpha} + \alpha(x - x_0)^2$$

and

$$f(x) = 2\{1 + \alpha(x - x_0)[\tan^{-1} [\alpha(x - x_0)] + \tan^{-1} (\alpha x_0)]\}$$

Furthermore, u has boundary values

$$u(0) = 0, \qquad u(1) = 0$$

The properties of u make it an ideal candidate for use in numerical studies of finite element solutions. To utilize this function, modify RMAT to read the values of x_0 and α, modify GETMAT to calculate $k(x)$ and $f(x)$, and modify EXACT to calculate $u(x)$ and $u'(x)$.

EXERCISES

3.8.6 Using a fairly smooth problem (say, $x_0 = 0.5$, $\alpha = 5$) and a uniform mesh of eight elements, study the effect of varying the order of integration on elements of different order. Run the complete set of problems possible; that is, for linear, quadratic, and cubic elements, use NINT $= 1, 2, 3$, and 4. Do not be unduly disturbed if some failures are observed. Observe the accuracy of the finite element solution as represented by the energy norm. Report and discuss your results and draw conclusions. The reader can modify this version of CODE1 so as to always use the best value of NINT for a particular element. What does "best" mean? Should computer time be considered when making this decision?

3.8.7 (a) Using the problem and mesh of Exercise 3.8.6, study *pointwise* accuracy of the finite element solution u_h and the flux σ_h by evaluating the solution at several points (say, 10 equally-spaced points) in an element or two near x_0. For each kind of element, the reader should discover certain points at which u_h tends to be most accurate and certain other points at which σ_h is most accurate.

(b) Extend the study to "rougher" problems (say, $\alpha = 50$). Report your results in the form of plots of error versus element coordinate ξ for the different kinds of elements. Draw conclusions.

The existence of these special points at which extra good accuracy is usually found is an extremely valuable aspect of finite element analysis. As mentioned in Chapter 1, this phenomenon of accuracy greater than average at these special points is called *superconvergence*. Applications codes typically evaluate the solution only at these points. You may wish to modify your version of CODE1 to automatically select the best output points for u_h and σ_h.

3.8.8 For both a "smooth" problem (small α) and a "rough" problem (large α), study the convergence rate of linear, quadratic, and cubic elements. Use uniform meshes of 2, 4, 8, and 16 elements ($h = \frac{1}{2}, \frac{1}{4}, \frac{1}{8}$, and $\frac{1}{16}$). Study the convergence in both the least-squares and energy norms. Plot $\log \|u - u_h\|$

versus $\log h$, as in Fig. 1.11, and determine the experimental convergence rates. Compare experimental values with the theoretical values from (2.6.12). Error estimates of the form (2.6.12) are asymptotic; that is, they exhibit increasing accuracy as h gets small, holding exactly in the limit as $h \longrightarrow 0$. It is interesting to ask how small h must be before asymptotic behavior (straight lines on the log error vs. $\log h$ plots) is observed. What do the results of your numerical experiments suggest?

3.8.9 For a "rough" problem (say, $x_0 = 0.75$ and $\alpha - 50$) and a limited number of nodal points (say, 30), try varying the spacing and order of the elements to obtain a highly accurate finite element solution (measured in the energy norm). Either extremely good luck or several trials will be required to achieve good results.

   A reasonable strategy is to start with a uniform grid, evaluate the solution at several points within each element and plot u versus x. From this plot, the regions in which rapid changes are taking place can be easily detected. According to our discussion of error estimates in Section 2.6, we should expect linear elements to behave poorly when d^2u/dx^2 is large, so that either h must be small or higher-order elements used (or both). Similarly, large values of d^3u/dx^3 give rise to large errors with quadratic elements, and so on. Use these ideas in constructing subsequent trial meshes.

   This exercise illustrates an important aspect of practical finite element analyses: uniform grids are rarely the most effective. The finite element method offers the opportunity to put small elements where they are needed and to use large elements in regions where they suffice. Of course, in practical applications, the exact solution is not available, but a preliminary finite element analysis using a uniform and coarse mesh can be used to obtain a rough picture of the solution. The mesh can then be refined in regions of rapid change in subsequent runs. The automation of this procedure is the subject of considerable research effort.

3.8.3 Applications Using Code 1

Heat Flow in Cylinders: The radial axisymmetric distribution of temperature in a long circular cylinder under steady state conditions is governed by the balance law

$$\sigma' = xf$$

and the constitutive equation (Fourier's Law)

$$\sigma = -kxu',$$

where x is the radial coordinate, u is the temperature, σ is the heat flux in the radial direction, f is the rate at which heat is generated in a unit area of the cross section, and k is the thermal conductivity of the material. Boundary conditions at the outer radius of the cylinder x_o may be any one of the following:

Temperature B.C. $u(x_o) = u_o$

Flux B.C. $\sigma(x_o) = \sigma_o$

Convective B.C. $\sigma(x_o) = p_o(u(x_o) - u_o)$

where u_o, σ_o, and p_o are given values of temperature, heat flux out of the cylinder and heat transfer coefficient. If the cylinder is hollow, with inner radius x_i, any of these same boundary condition types can be imposed on the inner boundary. In the boundary conditions on the inner boundary the sign of σ must be changed in order to maintain the interpretation of flux into the body. If the cylinder is solid, the boundary condition at the center, $x = 0$, must be that of zero heat flux, that is $\sigma(0) = 0$.

EXERCISES

3.8.10 Show that the boundary-value problem, in terms of temperature, is self adjoint (See Section 2.5).

3.8.11 Construct a variational statement of the boundary-value problem with convective boundary conditions as natural boundary conditions.

3.8.12 For a hollow cylinder with no heat generation ($f = 0$) and with inner boundary held at temperature u_i and outer boundary held at temperature u_o the temperature distribution is

$$u(x) = u_o + \frac{(u_i - u_o)\ln(x_o/x)}{\ln(x_o/x_i)}$$

Modify subroutines COEF and EXACT so that this boundary-value problem can be studied. For a thick-walled cylinder ($x_i = 1$ and $x_o = 10$) with $u_i = 100°$ and $u_o = 0°$, find by numerical experiment the number of equally spaced quadratic elements required so that all nodal point temperatures are calculated to an accuracy of $1°$ and investigate the use of unequal spacing of elements to obtain better computational efficiency.

Since no heat is generated in the material the heat flux should be the same across both inner and outer boundaries. The value of the flux is

$$\sigma = \frac{k(u_i - u_o)}{\ln(x_o/x_i)}$$

Compare the accuracy of flux calculation with that of temperature calculations.

3.8.13 In nuclear power reactors the fuel (uranium dioxide) is in the form of cylindrical pellets contained in long metal (zirconium alloy) tubes. The nuclear reaction causes heat to be generated in the fuel. This heat is then conducted through the wall of the metal tube into a cooling fluid (water). The heated water is used to produce steam that is supplied to turbines which turn electric generators.

Consider a model of a fuel rod in which the pellets have a radius $\bar{x}$ of .095 m and in which the wall thickness is .012 m. Let the conductivity of the fuel be 3 watts/m/°K and that of the tube 17 watts/m/°K. Assume that the heat generation in the fuel is given by $f(x) = 7.6 \times 10^8(1 - (x/\bar{x})^2)$ *watts*/m^3. Let the cooling water temperature be $610°K$ and the heat transfer coefficient between the tube and water be 3.5×10^5 *watts*/m^2/°K.

Modify subroutine COEF to accommodate the given data. Establish, by numerical experiment, a grid sufficient to calculate the nodal point temperatures to the nearest 10°. Calculate the maximum temperature in the tube wall.

Note that the strength of the tube decreases with temperature so that prediction of this temperature has a direct bearing on reactor safety.

Elastic Deformation of Cylinders: The problem of calculating the stresses and deformations in a body of revolution with axisymmetric loadings can be reduced to a one-dimensional problem in the case of a long circular cylinder (plane strain) or a thin circular disk (plane stress). Because of the symmetry in such problems all quantities of interest are functions of only the radial coordinate x. The quantities of interest are the radial displacement u, the radial stress σ_x and the tangential (or hoop) stress σ_θ.

The stresses are related to the displacement through the constitutive equation (Hooke's law)

$$\sigma_x = c_1 u' + c_2 u/x - c_3 e_T$$
$$\sigma_\theta = c_1 u/x + c_2 u' - c_3 e_T$$

where c_1, c_2, and c_3 are elastic constants and e_T is the strain caused by changes in temperature. The constants are given in terms of the usual engineering constants E (modulus of elasticity) and v (Poisson's ratio) by the following formulae:

Plane Strain	*Plane Stress*
$c_1 = E(1 - v)/(1 + v)(1 - 2v)$	$c_1 = E/(1 - v^2)$
$c_2 = Ev/(1 + v)(1 - 2v)$	$c_2 = Ev/(1 - v^2)$
$c_3 = E/(1 - 2v)$	$c_3 = E/(1 - v)$

The thermal strain e_T is the product of the coefficient of thermal expansion, α, for the material and the change in temperature, ΔT, from a stress free reference temperature. Loadings that are common include internal or external pressurization and centrifugal body forces due to spinning about the axis of symmetry.

Boundary conditions may be either essential (displacement) or natural (radial stress, σ_x). For hollow cylinders or disks with central holes, both the inner and outer boundaries may be given either kind of boundary condition. For solid bodies the displacement must be specified as zero at the center, i.e. $u = 0$ at $x = 0$.

The variational statement of the boundary value problem is to find $u = u(x)$ in H^1 so that

$$\int_{x_i}^{x_o} [c_1 x u'v' + c_2(u'v + uv') + c_1 uv/x]\, dx$$

$$= \int_{x_i}^{x_o} (vf_1 + v'f_2)\, dx + \sigma_x^o v(x_o) - \sigma_x^i v(x_i)$$

for all $v \in H^1$.

where f_1 and f_2 are distributed loading terms due to temperature and spin and σ_x^i and σ_x^o are the prescribed values of σ_x at the inner and outer boundaries. The loading terms are

$$f_1 = \rho\omega^2 x^2 + c_3 \alpha \Delta T$$

$$f_2 = c_3 \alpha \Delta T x$$

where ρ is the mass density of the material and ω is the angular velocity of the body.

EXERCISES

3.8.14   Make the necessary modifications to CODE1 to solve the boundary-value problems described in this subsection. These modifications consist of

(a) modifications of subroutine RMAT to read values of E, v, α, ΔT, ρ, and ω,

(b) modifidations of subroutine GETMAT to calculate and return to ELEM values of the coefficients and load functions appearing in the variational statement of the boundary-value problem,

(c) modifications of subroutine ELEM (statements 11 and 12) to calculate the element stiffness matrix and load vector according to the variational statement, and

(d) modifications to subroutine EXACT to calculate displacement, u, and stress, σ_x and σ_θ, for problems with known solutions (as in Exercises 3.8.15 and 3.8.16).

3.8.15   For an internally pressurized cylinder ($\Delta T = \omega = 0$) in plane strain the displacement is given by

$$u = \frac{x_i^2 \sigma_x^i (1 + v)}{E(x_o^2 - x_i^2)} \left[(1 - 2v)x + \frac{x_o^2}{x} \right]$$

Consider a steel pressure vessel ($E = 30 \times 10^6$ psi, $v = 0.3$) with $x_i = 1$ inch, $x_o = 4$ inches, and $\sigma_x^i = 8000$ psi. Use three quadratic elements to do a stress analysis. Plot exact and finite element values of σ_x and σ_θ as functions of x. The yielding of the material depends on the principal stress difference $\sigma^* = \sigma_\theta - \sigma_x$. How accurately does the three element model predict σ^*? If the three element model is not satisfactory, try a better one. Is nonuniform grid spacing indicated for this problem?

3.8.16   For a thin solid disk in plane stress, the displacement caused by spinning is

$$u = \frac{\rho \omega^2 x}{8E} [(1 - v)(3 + v)x_o^2 - (1 - v^2)x^2]$$

Analyze a solid steel turbine wheel ($\rho = 7.32 \times 10^{-4}$ lb sec²/in.) of radius 12 inches rotating at 10,000 rpm. Use five quadratic elements and plot σ_{xh} and $\sigma_{\theta h}$ as functions of x. (Remember that for a solid disk there must be an essential boundary condition, $u(0) = 0$.)

   Analyze the same disk with a very small hole in the center by removing the essential boundary condition. Note the large increase in σ_θ at $x = 0$. The exact value of σ_θ for a hole of zero radius is twice the value for a solid disk. Discuss the calculation of maximum values of stress in your finite element solution.

3.8.17   Consider a long hollow cylindrical chemical reactor pressure vessel. Let the dimensions and pressure be the same as those of Exercise 3.8.15, but suppose that, in addition to the pressure loading, the vessel is subject to a temperature rise $\Delta T = \Delta T_o + \frac{\Delta T_i - \Delta T_o}{\ln (x_o/x_i)} \ln (x_o/x)$ with $\Delta T_i = 250°F$ and $\Delta T_o = 50°F$.

   Modify GETMAT to provide the loading terms appropriate to this problem and use a suitable grid to perform a stress analysis of the vessel. Plot σ^* and compare with the results of Exercise 3.8.15. (Use $\alpha = 6.5 \times 10^{-6}$ in/in°F for steel.) Note that thermal stress analysis often requires the calculation of unknown temperature distributions in order that the loads be known. Finite element codes often use the same grids to perform both calculations. The nodal-point temperatures from the thermal analysis are saved and used as data for the stress analysis.

3.8.18   Rotors are often shrink-fitted to shafts by heating the rotor, inserting the shaft, and then allowing the assembly to cool. When the heated part has a uniform (but higher) temperature, there is no stress so that temperature changes can be measured from this reference temperature. Consider a steel rotor shrunk onto a solid copper-berylium shaft. Assume plane stress conditions and suppose that there is a perfect fit (no stress and no gap) when the shaft is inserted into the hole in the rotor. Model the problem with five quadratic elements in the shaft and ten quadratic elements in the rotor. The properties to be used are given below

	Shaft	*Rotor*
Inner radius x_i	0 inches	2 inches
Outer radius x_o	2 inches	20 inches
Modulus of Elasticity E	17×10^6 psi	30×10^6 psi
Poisson's Ratio ν	0.3	0.3

	Shaft	*Rotor*
Coefficient of Thermal Expansion α	0	6×10^{-6} $in/in/°F$
Temperature Change	0	$200°F$
Mass Density ρ	7.71×10^{-4} $lb\ sec^2/in^4$	7.32×10^{-4} $lb\ sec^2/in^4$

Calculate and plot the stresses due to shrink fitting. Note that the hoop stress σ_θ is tensile (positive) in the rotor and compressive (negative) in the shaft and has a discontinuity at the interface $x = 2$ while the displacement and radial stress are continuous.

Calculate and plot the stresses induced by spinning the rotor and shaft. How fast can the assembly rotate before the rotor becomes loose (that is, before the total radial stress at the interface becomes tensile)?

4

TWO-DIMENSIONAL PROBLEMS

4.1 INTRODUCTION

The principal ingredients of the finite element method for constructing approximate solutions of problems are:

1. The formulation of the problem in a variational framework in which the appropriate space H of admissible functions is identified.

2. The construction of a finite element mesh and piecewise-polynomial basis functions defined on the mesh, which generate a finite-dimensional subspace of H.

3. The construction of an approximation of the variational boundary-value problem on a finite element subspace H^h of H. This entails the calculation of element matrices and the generation of a sparse system of linear algebraic equations in the values of the approximate solution at nodal points in the mesh.

4. The solution of the algebraic system.

5. The examination of the characteristics of the approximate solution and, if possible, an estimation of the inherent approximation error.

These steps form the basis of most finite element methods for not only one-dimensional problems, but, more importantly, for boundary-value problems in two and three dimensions.

Our objective here is to develop, in a logical manner, the natural extension of the earlier developments to two dimensions. The governing equation for

131

the boundary-value problem now becomes a partial differential equation which is to be satisfied by the solution inside some two-dimensional domain Ω. Boundary data are thus specified on the curve defining the boundary of Ω. Instead of line elements, the finite elements now assume simple two-dimensional shapes, such as triangles and quadrilaterals, and these elements fit together to make up the discretization (the mesh) approximating the domain Ω of the solution u. The inherent ability of such elements to represent domains with very irregular geometries underlies the practical value of the method for the approximate solution of difficult boundary-value problems in numerous research and industrial applications.

In this chapter we consider problems for which u is a scalar-valued function of position. Among physical phenomena modeled by such problems are heat conduction (u being the temperature), seepage flow (u being the hydraulic head), and the transverse deflection of an elastic membrane (u being the deflection). In other problems, such as stress analysis and flow of viscous fluids, the state variable u is a vector-valued function, sometimes called a *vector field*, such as displacement or velocity. The extension of the techniques described in this chapter to problems with vector-valued solutions is discussed briefly in Chapter 6.

4.2 TWO-DIMENSIONAL BOUNDARY-VALUE PROBLEMS

The development of boundary-value problems describing physical phenomena in two dimensions follows closely the one-dimensional treatment given in Chapter 2, differing only in aspects dictated by the higher dimensionality. In this section we sketch the development of linear, elliptic, self-adjoint second-order boundary-value problems, based on classical conservation principles.

4.2.1 Some Preliminaries

The domain $\bar{\Omega}$ of our problem is composed of two parts, the interior Ω and the boundary $\partial\Omega$. We consider only *bounded* domains (i.e., no part of Ω extends to infinity in any direction), with reasonably smooth boundaries.* In general, the boundary can be defined by the parametric equations $x = x(s)$

---

* See, for example, Volume IV of this series for a discussion of smoothness conditions on $\partial\Omega$. We confine our attention here to regions smooth enough for the direction of a unit vector **n** drawn perpendicular to the boundary to be a continuous function of position along the boundary except at corners with interior angles α, $0 < \alpha < \pi$. Domains with re-entrant corners ($\alpha > \pi$) give rise to singular solutions. Such problems are discussed in Volume II of this series.

and $y = y(s)$, where s is the arc length along $\partial\Omega$ measured from some arbitrary reference point. When referring to the values of a function, say g, which is defined at points on the boundary, we will write $g(s) \equiv g(x(s), y(s))$ for s in $\partial\Omega$.

The primary dependent variable in our problem is the state variable $u = u(x, y)$. As an essential condition, we require that u be a smooth function in Ω. The degree of smoothness we require depends on the data of the problem, including the functions $x(s)$ and $y(s)$ that define $\partial\Omega$. We shall assume throughout this section that, unless otherwise noted, all functions are smooth enough for the operations we perform to be valid.

Our physical statement of the problem will contain expressions involving the rate of change with respect to position in Ω of the scalar field u. In general, the value of u at a point (x, y) changes at different rates as we move from (x, y) in different directions. The rate of change of u at a point (x, y) is defined by a vector-valued function, denoted ∇u, called the *gradient* of u. If **i** and **j** denote unit vectors directed along the x- and y-axis, respectively, then the components of ∇u at a point (x, y) with respect to these basis vectors are $\partial u(x, y)/\partial x$ and $\partial u(x, y)/\partial y$, so that

$$\nabla u(x, y) = \frac{\partial u(x, y)}{\partial x}\mathbf{i} + \frac{\partial u(x, y)}{\partial y}\mathbf{j} \qquad (4.2.1)$$

Note that (4.2.1) can be interpreted as the construction of a vector field ∇u by operating on u with the vector differential operator

$$\nabla = \frac{\partial}{\partial x}\mathbf{i} + \frac{\partial}{\partial y}\mathbf{j} \qquad (4.2.2)$$

The gradient $\nabla u(x, y)$ determines the total rate of change of u at (x, y) in any direction. In particular, let **t** be a unit vector that makes an angle θ with the positive x-axis. Then **t** has components $\cos\theta$ and $\sin\theta$, so that $\mathbf{t} = \cos\theta\mathbf{i} + \sin\theta\mathbf{j}$. The rate at which u changes as one moves from a point (x, y) in the direction of **t** is called the *directional derivative* of u with respect to **t** and is written $du(x, y)/dt$. The directional derivative is calculated according to

$$\frac{du(x, y)}{dt} = \nabla u(x, y) \cdot \mathbf{t} = \frac{\partial u(x, y)}{\partial x}\cos\theta + \frac{\partial u(x, y)}{\partial y}\sin\theta \qquad (4.2.3)$$

The second quantity of physical interest in our boundary-value problems is the flux **σ**. The flux, like the gradient of u, is a vector-valued function or *vector field*. In Fig. 4.1a, a flux field is represented schematically as vectors that vary in magnitude and direction within $\bar{\Omega}$. The flux vector $\boldsymbol{\sigma}(s)$ at the point s on the boundary is shown in Fig. 4.1b. The flux crossing the boundary

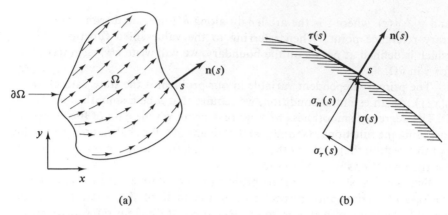

FIGURE 4.1 *(a) Representation of the vector-valued function* $\boldsymbol{\sigma} = \boldsymbol{\sigma}(x, y)$*; (b) resolution of the flux* $\boldsymbol{\sigma}(s)$ *at a point s on* $\partial\Omega$ *into normal flux* $\sigma_n(s)$ *and tangential flux* $\sigma_\tau(s)$ *components.*

at s is given by the component of $\boldsymbol{\sigma}(s)$ in the direction of a unit outward normal $\mathbf{n}(s)$ to $\partial\Omega$:

$$\sigma_n(s) = \boldsymbol{\sigma}(s) \cdot \mathbf{n}(s) \tag{4.2.4}$$

The component in the direction of the unit tangent $\boldsymbol{\tau}(s)$ is $\boldsymbol{\sigma}(s) \cdot \boldsymbol{\tau}(s)$, as indicated in the figure.

Consider an arbitrary subregion of Ω, say ω, containing the point P_0 whose coordinates are (x_0, y_0). Figure 4.2a shows the distribution of normal flux $\sigma_n(s)$ across the boundary $\partial\omega$. The net (total) flux crossing the boundary is

$$\Sigma_\omega \equiv \int_{\partial\omega} \sigma_n(s)\, ds \tag{4.2.5}$$

If we divide Σ_ω by the area A_ω of the subregion, we can view the result as the average value of the amount of $\boldsymbol{\sigma}$ flowing into ω per unit area. The limit of this ratio as ω decreases in size, always containing the point P_0, is called the *divergence* of the flux at P_0 and is often abbreviated div $\boldsymbol{\sigma}(x_0, y_0)$. Taking as ω the square subregion containing P_0 shown in Fig 4.2b, we calculate $\Sigma_\omega = \Delta\sigma_x\,\Delta y + \Delta\sigma_y\,\Delta x$. Then, using the mean-value theorem to expand components σ_x and σ_y of $\boldsymbol{\sigma}$ about point P_0, as indicated in the figure, we divide by the area $\Delta x\,\Delta y$ and calculate the limit as Δx, Δy go to zero. In this way, we obtain the formula for the divergence of the vector field $\boldsymbol{\sigma}$ at the point (x, y),

$$\text{div }\boldsymbol{\sigma}(x, y) = \frac{\partial\sigma_x(x, y)}{\partial x} + \frac{\partial\sigma_y(x, y)}{\partial y} \tag{4.2.6}$$

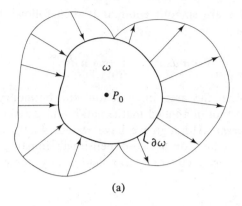

(a)

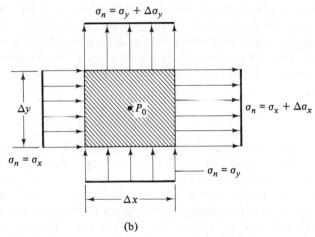

(b)

FIGURE 4.2 (a) *Distribution of normal flux on boundary $\partial\omega$ of subregion ω; (b) magnitude of normal flux on boundary of square region.*

Note that by using the vector operator $\mathbf{V}$ defined in (4.2.2), the divergence of $\boldsymbol{\sigma}$ can also be written

$$\text{div } \boldsymbol{\sigma} = \mathbf{V} \cdot \boldsymbol{\sigma} \qquad (4.2.7)$$

Recall the definition of $\mathbf{V} \cdot \boldsymbol{\sigma}$ as the density of the net flux per unit area at a point. It follows that the total flux Σ out of the region Ω can be written*

$$\Sigma = \int_{\Omega} \mathbf{V} \cdot \boldsymbol{\sigma} \, dx \, dy \qquad (4.2.8)$$

---

* Here and hereafter, we use the notation $\iint_{\Omega} dx \, dy = \int_{\Omega} dx \, dy$.

provided that Ω and $\boldsymbol{\sigma}$ are smooth enough. It then follows from (4.2.4), (4.2.5), and (4.2.8) that

$$\int_\Omega \nabla \cdot \boldsymbol{\sigma} \, dx \, dy = \int_{\partial\Omega} \boldsymbol{\sigma} \cdot \mathbf{n} \, ds \qquad (4.2.9)$$

The relation between the area integral and the boundary integral in (4.2.9) is an important tool in applied mathematics and is referred to as the *Gauss divergence theorem*. Although we have stated (4.2.9) in terms of a special vector field $\boldsymbol{\sigma}$ defined on a two-dimensional domain, the result is generally the same for any vector or tensor field in any number of dimensions.

4.2.2   Physical Principles

The physical situations we wish to consider are governed by a linear constitutive law and a conservation principle. The linear constitutive equation in our physical problem establishes that, at each point in the body, the flux is proportional to the gradient of the state variable. The factor of proportionality is denoted k and is referred to as the material modulus (coefficient or property), as in Chapter 2. Thus,*

$$\boldsymbol{\sigma}(x, y) = -k(x, y) \, \nabla u(x, y) \qquad (4.2.10)$$

Clearly, different materials will be characterized by different material moduli, $k = k(x, y)$. As in Chapter 2, we will always assume that $|k(x, y)| > k_0 = \text{constant} > 0$, throughout $\bar{\Omega}$.

The conservation principle (or balance law) states that within any portion of the domain, the net flux across the boundary of that part must equal the total quantity produced by internal sources. The mathematical implications of the balance law take different forms for different parts of the domain.

Let us next examine the implications of the conservation principle. Consider, for example, a portion ω of material surrounding a point P_0 inside Ω (Fig. 4.3), in which the material properties are smooth. The balance law defined earlier establishes that the net flux across the boundary $\partial\omega$, given by (4.2.5), must be balanced by the total quantity supplied by sources within ω. If f denotes the source per unit area, then we must have

$$\int_{\partial\omega} \boldsymbol{\sigma} \cdot \mathbf{n} \, ds = \int_\omega f \, dx \, dy \qquad (4.2.11)$$

---

* The form of (4.2.10) implies that, at each point (x, y), the direction of the flux is parallel to the direction of $\nabla u(x, y)$; in other words, the material is *isotropic*. The formulation and finite element treatment of problems involving *anisotropic* materials are essentially the same as those given for the isotropic case.

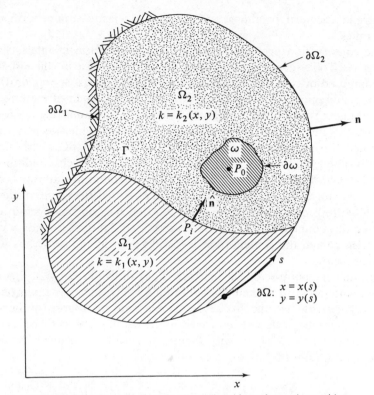

FIGURE 4.3 *Domain of a two-dimensional boundary-value problem.*

Using the divergence theorem (4.2.9) to transform the surface integral in (4.2.11) to an area integral, we have

$$\int_{\omega} (\nabla \cdot \boldsymbol{\sigma} - f)\, dx\, dy = 0 \tag{4.2.12}$$

for all subregions ω in Ω. Since ω is an arbitrary region in which $\nabla \cdot \boldsymbol{\sigma}$ and f are smooth, the integrand in (4.2.12) must be zero at all points interior to ω. Thus, within such "smooth" regions the local form of the balance law is

$$\nabla \cdot \boldsymbol{\sigma}(x, y) = f(x, y) \tag{4.2.13}$$

For completeness, we suppose that, in addition to f, there may exist internal sources with an intensity proportional to u. Letting the proportionality factor be $-b(x, y)$, (4.2.13) becomes

$$\nabla \cdot \boldsymbol{\sigma}(x, y) + b(x, y)u(x, y) = f(x, y) \tag{4.2.14}$$

Examples of sources proportional to u include temperature-dependent exothermic chemical reactions and distributed elastic supports of elastic membranes.

The conservation principle takes on a different form at interfaces and boundaries. To fix ideas, let us consider the particular case in which the body $\bar{\Omega}$ is composed of two distinct materials, one occupying a subregion Ω_1 and the other subregion Ω_2, as shown in Fig. 4.3. Within each of these regions, the modulus k is assumed to be given by smooth functions (e.g., constants) k_1 and k_2, as indicated. The curve defining the interface between Ω_1 and Ω_2 is denoted Γ. Similarly, we suppose that the boundary $\partial\Omega$ of the body is naturally divided into two parts, $\partial\Omega_1$ and $\partial\Omega_2$, on which conditions are imposed which characterize the effect of the surrounding exterior medium on the behavior of the body. On $\partial\Omega_1$, we suppose that the values of the state variable u are prescribed as $u(s) = \hat{u}(s)$, so that it is on this portion of $\partial\Omega$ that essential boundary conditions are prescribed. Natural boundary conditions arising from the conservation law will be specified on $\partial\Omega_2$, as will be explained below.

Consider the point P_i on the interface Γ, as shown in Fig. 4.3. Figure 4.4 shows a material strip of the body containing P_i. This strip is assumed to be sufficiently narrow that the flux across its ends and the source (proportional to the area) can be neglected compared to the net flux across the sides of the strip. As the thickness of the strip shrinks to zero, the balance law for this strip assumes the form

$$\Sigma = \int_{s_1}^{s_2} (-\boldsymbol{\sigma}^{(-)} \cdot \mathbf{n} + \boldsymbol{\sigma}^{(+)} \cdot \mathbf{n}) \, ds = 0$$

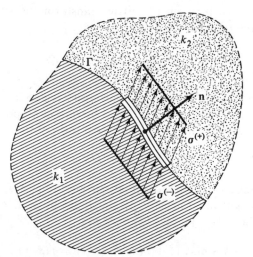

FIGURE 4.4 *A strip of the domain containing a portion of the interface Γ in the neighborhood of P_i.*

where s_1 and s_2 are the endpoints of the strip. Because the region of integration is arbitrary, the local balance law at points on the interface reduces to conditions on the jump in $\boldsymbol{\sigma} \cdot \mathbf{n} = \sigma_n$ across Γ:

$$[\![\sigma_n(s)]\!] = \sigma_n^{(+)}(s) - \sigma_n^{(-)}(s) = 0, \qquad s \in \Gamma \qquad (4.2.15)$$

We now turn to the boundary conditions on $\partial\Omega_2$. A region containing a typical boundary point P_b is shown in Fig. 4.5. The value of the normal flux

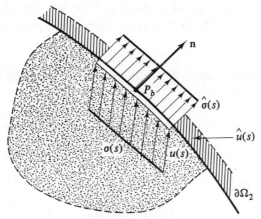

FIGURE 4.5 *A strip of the domain containing a portion of the boundary $\partial\Omega_2$ in the neighborhood of point* P$_b$.

$\hat{\sigma}(s)$ through the surrounding material immediately adjacent to the boundary is assumed to be proportional to the difference between the value $u(s)$ on the boundary and the given value $\hat{u}(s)$ in the exterior medium. Thus, if $p(s)$ is the factor of proportionality at s,

$$\hat{\sigma}(s) \equiv p(s)[u(s) - \hat{u}(s)]$$

Balancing the flux in the strip containing P_b gives

$$\sigma_n(s) \equiv \boldsymbol{\sigma}(s) \cdot \mathbf{n}(s) = \hat{\sigma}(s) \qquad (4.2.16)$$

Hence,

$$\sigma_n(s) = p(s)[u(s) - \hat{u}(s)], \qquad s \in \partial\Omega_2 \qquad (4.2.17)$$

The physical situation often dictates other forms of the boundary condition, which can be viewed as special cases of (4.2.16) or (4.2.17). For example, we recover the essential boundary condition on u from (4.2.17) by taking the limit as $p(s) \rightarrow \infty$ while $\sigma_n(s)$ remains bounded. This procedure corresponds to a two-dimensional version of the penalty method discussed in Section 3.8.1. Boundary conditions (4.2.16) and (4.2.17), which are consequences of the conservation principle, are called *natural* boundary conditions. Boundary conditions that prescribe the state variable are *essential* boundary conditions.

4.2.3　Statement of the Boundary-Value Problem

　　The final mathematical statement of our boundary-value problem is obtained by eliminating $\boldsymbol{\sigma}$ and σ_n from (4.2.14) through (4.2.17) using the constitutive equation (4.2.10). The data defining the problem consist of the following:

1. The boundaries $\partial\Omega_1$, $\partial\Omega_2$, and the interface Γ defined by the parametric equations

$$x = x(s), \quad y = y(s), \quad s \in \partial\Omega \quad \text{or} \quad s \in \Gamma$$

2. The source distribution $f = f(x, y)$ in Ω_i, $i = 1, 2$.

3. The material coefficients $k_i = k_i(x, y)$ and $b_i = b_i(x, y)$ for $(x, y) \in \Omega_i$, $i = 1, 2$.

4. The prescribed value $\hat{u}(s)$ for $s \in \partial\Omega_1$.

5. The value of the boundary coefficient $p(s)$ and $\hat{u}(s)$ on $\partial\Omega_2$ or the prescribed value $\hat{\sigma}(s)$ for $s \in \partial\Omega_2$.

　　Given the foregoing data, the problem is to find the function $u = u(x, y)$ that satisfies:

1. The governing partial differential equation at points interior to the smooth subdomains Ω_1 and Ω_2,

$$-\boldsymbol{\nabla} \cdot (k(x, y)\boldsymbol{\nabla}u(x, y)) + b(x, y)u(x, y) = f(x, y)$$
$$\text{for } (x, y) \in \Omega_i, i = 1, 2 \qquad (4.2.18a)$$

2. The jump condition at points on the interface Γ,

$$[\![k\boldsymbol{\nabla}u \cdot \mathbf{n}]\!] = 0, \quad s \in \Gamma \qquad (4.2.18b)$$

3. The essential boundary condition on $\partial\Omega_1$,

$$u(s) = \hat{u}(s), \quad s \in \partial\Omega_1 \qquad (4.2.18c)$$

4. The natural boundary condition on $\partial\Omega_2$,

$$-k(s)\frac{\partial u(s)}{\partial n} = p(s)[u(s) - \hat{u}(s)], \qquad s \in \partial\Omega_2$$

or

$$-k(s)\frac{\partial u(s)}{\partial n} = \hat{\sigma}(s), \qquad s \in \partial\Omega_2$$

(4.2.18d)

The special case of problem (4.2.18), in which $b \equiv 0$ and only natural boundary conditions of the form $-k(s)[\partial u(s)/\partial n] = \hat{\sigma}(s)$ are specified and $\partial\Omega_2 - \partial\Omega$, requires two qualifying remarks. First, we note that (4.2.18) then determines the solution u only to within an additive constant. The second point is that, in order for a solution u to exist at all, the data f and $\hat{\sigma}$ must satisfy a compatibility relation. This relation is simply a statement of the conservation principle for the entire domain Ω and requires that f and $\hat{\sigma}$ be such that

$$\int_\Omega f \, dx \, dy = \int_{\partial\Omega} \hat{\sigma} \, ds \tag{4.2.19}$$

The development and statement of the boundary-value problem have been made assuming that all functions were as smooth as necessary for the operations indicated to be valid. The only consideration made for lack of smoothness in any of the data was the treatment of the material interface Γ across which the coefficients k, b and the source f may be discontinuous.

The exact specification of smoothness requirements on the data and the relation of this smoothness to the properties (existence and smoothness) of the solution u is beyond the scope of this volume. In general, we can say, however, that the smoothness requirements for the statement of the boundary-value problem as given by (4.2.18) are much more restrictive than those required by the variational statement of the problem given in the following section.

EXERCISES

4.2.1 Let Ω consist of the region of the x, y-plane such that $x > 0$, $y > 0$, and $x^2 + y^2 < 1$.

(a) Sketch the region.

(b) Write the functions $x(s)$ and $y(s)$ that define each of the three parts of $\partial\Omega$, where s is the arc length along $\partial\Omega$ measured from the point $(1, 0)$ and proceeding counterclockwise.

(c) For each of the three parts of $\partial\Omega$, write the functions $n_x = n_x(s)$ and $n_y = n_y(s)$ that define the x- and y-components of the unit vector normal to $\partial\Omega$ (pointing outward from the interior of Ω).

4.2.2 Consider the function given by $g(x, y) = x^2 + y^2$ on a unit square domain $\bar{\Omega}: 0 \le x, y \le 1$.

(a) The contours (or level lines) of g are the loci of points at which $g(x, y)$ is constant. Draw the contours of g in Ω for values of $g(x, y) = 0.25$, 0.5, 0.75, 1, and 1.25. Note that the distance between contours is inversely proportional to the magnitude of ∇g.

(b) Calculate the components and magnitude of ∇g at the points $(0.5, 0)$, $(0.707, 0.707)$, $(1, 0.5)$, and $(0.2, 0.4)$. Draw these values on the contour plot of g. (The values of ∇g are vectors.)

(c) Calculate the rate of change of $g(x, y)$ at the point $(0.2, 0.4)$ in the directions given by the following vectors:

(i) $\begin{bmatrix} 1 \\ 0 \end{bmatrix}$ (ii) $\begin{bmatrix} 1 \\ 1 \end{bmatrix}$ (iii) $\begin{bmatrix} 2 \\ 1 \end{bmatrix}$

(d) Let the arc length s be measured counterclockwise from the point $(1, 0)$. Calculate the value of the normal derivative $\partial g/\partial n$ as a function of s along the portions of the boundary

(i) $0 < x < 1$, $y = 0$ $(3 < s < 4)$

(ii) $x = 1$, $0 < y < 1$ $(0 < s < 1)$

4.2.3 Consider the scalar field $g = g(x, y)$ of Exercise 4.2.2 defined on the region Ω of Exercise 4.2.1. Let the flux $\boldsymbol{\sigma}$ be given by $\boldsymbol{\sigma} = \nabla g$.

(a) Calculate the normal component of $\boldsymbol{\sigma}$ as a function of s along each part of the boundary $\partial\Omega_i$.

(b) Calculate the total flux across each $\partial\Omega_i$,

$$\Sigma_i = \int_{\partial\Omega_i} \sigma_n \, ds$$

(c) Calculate the divergence of the flux $\nabla \cdot \boldsymbol{\sigma}$ and the integral over Ω of $\nabla \cdot \boldsymbol{\sigma}$.

(d) Verify that the total flux across $\partial\Omega$ is equal to the integral over Ω of the divergence of $\boldsymbol{\sigma}$.

(e) Find $f(x, y)$, so that $\nabla \cdot (\nabla g) = f$ at every point in Ω.

4.2.4 When $k = 1$, $b = 0$, and $f = 0$, the partial differential equation (4.2.18a) is reduced to the form known as *Laplace's equation*.

(a) Show that Laplace's equation is

$$\frac{\partial^2 u}{\partial x^2} + \frac{\partial^2 u}{\partial y^2} = 0$$

(b)   Laplace's equation is often written as

$$\Delta u = 0$$

where Δ is a differential operator (called Laplace's operator or the Laplacian). Show that the definition of Δ as

$$\Delta \equiv \nabla \cdot \nabla$$

is consistent with the form of Laplace's equation given in part (a).

(c)   Let u and v be scalar functions defined on a region Ω and its boundary $\partial\Omega$.

   (i)   Show (by direct calculation) that

$$\nabla \cdot (v\nabla u) = v(\Delta u) + \nabla v \cdot \nabla u$$

   (ii)   Use the divergence theorem (4.2.9) to show that

$$-\int_{\Omega} v\,\Delta u\,dx\,dy = \int_{\Omega} \nabla v \cdot \nabla u\,dx\,dy - \int_{\partial\Omega} v\,\frac{\partial u}{\partial n}\,ds$$

[*NOTE:*   The equation in part (ii) is known as Green's formula or the Green–Gauss theorem for the Laplacian.]

4.2.5   Using the notation in Exercise 4.2.4, consider the problem of finding u such that

$$\Delta u(x, y) = 2, \qquad (x, y) \in \Omega$$
$$u(s) = 0, \qquad s \in \partial\Omega$$

where Ω is a circle of radius r_0 and the origin of our x, y-coordinate system is at the center of the circle. Determine the constant C and the radius r_0 such that the function

$$u(x, y) = C(x^2 + y^2 - 1)$$

is a solution to the problem.

4.2.6   Suppose that Ω is a rectangle with a side of length a parallel to the x-axis and a side of length b parallel to the y-axis, the origin being at the lower left corner.

(a)   Determine data $f(x, y)$ and $\hat{\sigma}(s)$ such that the function

$$u(x, y) = \tfrac{1}{6}x^3 y^3$$

is a solution of the problem

$$\Delta u(x, y) = f(x, y), \qquad (x, y) \in \Omega$$
$$\frac{\partial u(s)}{\partial n} = \hat{\sigma}(s), \qquad s \in \partial\Omega$$

(b) Verify that f and $\hat{\sigma}$ satisfy the compatibility condition (4.2.19).

(c) Let $\partial\Omega$ be divided into two parts:

$$\partial\Omega_1 = \text{the points } x = 0, 0 \le y \le b \quad \text{and} \quad y = 0, 0 \le x \le a$$
$$\partial\Omega_2 = \text{the points } x = a, 0 \le y \le b \quad \text{and} \quad y = b, 0 \le x \le a$$

Determine data f, $\hat{u}$, and $\hat{\sigma}$ such that the function u given in part (a) is a solution of the boundary-value problem

$$\Delta u(x, y) = f(x, y), \qquad (x, y) \in \Omega$$
$$u(s) = \hat{u}(s), \qquad s \in \partial\Omega_1$$
$$\frac{\partial u(s)}{\partial n} = \hat{\sigma}(s), \qquad s \in \partial\Omega_2$$

4.3 VARIATIONAL BOUNDARY-VALUE PROBLEMS

The construction of our variational formulation of the boundary-value problem (4.2.18) begins, as usual, by defining the residual r:

$$r(x, y) = -\mathbf{\nabla} \cdot [k(x, y)\mathbf{\nabla}u(x, y)] + b(x, y)u(x, y) - f(x, y)$$

To "test" the residual over arbitrary subregions, we multiply r by a sufficiently smooth test function v, integrate over each domain in which rv is smooth, and set the resulting weighted average of r equal to zero. For the problem whose domain is shown in Fig. 4.3, we must integrate separately over Ω_1 and Ω_2 since the second derivatives of u are not integrable along the interface Γ. This procedure gives

$$\int_{\Omega_1} [-\mathbf{\nabla} \cdot (k\mathbf{\nabla}u) + bu - f]v \, dx \, dy + \int_{\Omega_2} [-\mathbf{\nabla} \cdot (k\mathbf{\nabla}u) \qquad (4.3.1)$$
$$+ bu - f]v \, dx \, dy = 0$$

A two-dimensional "integration-by-parts formula" is needed to reduce the first term in each of these integrals to terms containing only first derivatives. By the product rule for differentiation, we find (see Exercise 4.3.1) that

$$\mathbf{\nabla} \cdot (vk \, \mathbf{\nabla}u) = k \, \mathbf{\nabla}u \cdot \mathbf{\nabla}v + v\mathbf{\nabla} \cdot (k \, \mathbf{\nabla}u) \quad \Big\}$$

or $\qquad\qquad\qquad\qquad\qquad\qquad\qquad\qquad\qquad\qquad\qquad\qquad$ (4.3.2)

$$v\mathbf{\nabla} \cdot (k \, \mathbf{\nabla}u) = \mathbf{\nabla} \cdot (vk \, \mathbf{\nabla}u) - k \, \mathbf{\nabla}u \cdot \mathbf{\nabla}v \quad \Big\}$$

Substitution of (4.3.2) into (4.3.1) yields

$$\int_{\Omega_1} (k\, \nabla u \cdot \nabla v + buv - fv)\, dx\, dy + \int_{\Omega_2} (k\, \nabla u \cdot \nabla v + buv$$

$$- fv)\, dx\, dy - \int_{\Omega_1} \nabla \cdot (vk\, \nabla u)\, dx\, dy \qquad (4.3.3)$$

$$- \int_{\Omega_2} \nabla \cdot (vk\, \nabla u)\, dx\, dy = 0$$

The last two integrals in (4.3.3) can be transformed into boundary integrals using the divergence theorem (4.2.9) with $(vk\, \nabla u)$ used in place of $\boldsymbol{\sigma}$. We obtain

$$-\int_{\Omega_1} \nabla \cdot (vk\, \nabla u)\, dx\, dy - \int_{\Omega_2} \nabla \cdot (vk\, \nabla u)\, dx\, dy$$

$$= -\int_{\partial(\Omega_1)} k\, \frac{\partial u}{\partial n} v\, ds - \int_{\partial(\Omega_2)} k\, \frac{\partial u}{\partial n} v\, ds \qquad (4.3.4)$$

where $\partial(\Omega_1)$ and $\partial(\Omega_2)$ are the boundaries of subregions Ω_1 and Ω_2, the direction of integration is counterclockwise in each of Ω_1 and Ω_2, and $\partial u / \partial n = \nabla u \cdot \mathbf{n}$. We must be careful to identify the functions in (4.3.4) with the domains on which they are defined. In Fig. 4.6, which shows Ω_1 and Ω_2 separated for clarity, the boundary of each domain is divided into two parts—the parts of $\partial(\Omega_i)$ that do not coincide with the interface Γ are denoted $\partial(\Omega_i) - \Gamma$, $i = 1, 2$. We decompose each of the boundary integrals in (4.3.4) into two corresponding parts, obtaining

$$-\int_{\partial(\Omega_1)-\Gamma} k\, \frac{\partial u}{\partial n} v\, ds - \int_{\partial(\Omega_2)-\Gamma} k\, \frac{\partial u}{\partial n} v\, ds$$

$$+ \int_{\Gamma} \left(-k\, \frac{\partial u}{\partial n}\right)_1 v\, ds + \int_{\Gamma} \left(-k\, \frac{\partial u}{\partial n}\right)_2 v\, ds \qquad (4.3.5)$$

wherein the notation

$$\left(-k\, \frac{\partial u}{\partial n}\right)_i \quad \text{indicates that} \quad \left(-k\, \frac{\partial u}{\partial n}\right)$$

is to be evaluated on region i. Noting that the outward normal $\mathbf{n}_1$ to region Ω_1 is the negative of $\mathbf{n}_2$ at each point on Γ, we rewrite the sum of the last two integrals in (4.3.5) as

$$\int_{\Gamma} \left[-k^{(+)}\, \frac{\partial u^{(+)}}{\partial n} + k^{(-)}\, \frac{\partial u^{(-)}}{\partial n}\right] v\, ds \qquad (4.3.6)$$

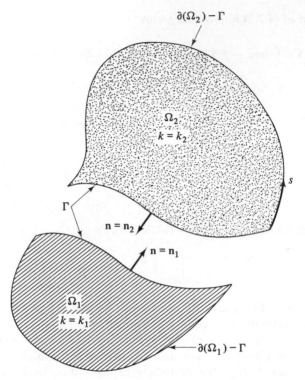

FIGURE 4.6 *Decomposition of regions of integration for boundary integrals around $\partial(\Omega_1)$ and $\partial(\Omega_2)$.*

where we have used the notation introduced in the preceding section. We recognize that the integrand in (4.3.6) is exactly $v[\![\sigma_n(s)]\!]$, which, according to (4.2.18b), is zero. Hence, the integral in (4.3.6) vanishes.

Returning to (4.3.5), we note that the first two integrals can be combined into a single integral over the entire boundary $\partial\Omega$. We also note that the integrands of the first two integrals in (4.3.3) contain at most first derivatives of u and v, so, if u and v are smooth enough, these integrals can be combined into a single integral over the entire domain Ω. The result is

$$\int_\Omega (k\,\nabla u \cdot \nabla v + buv - fv)\,dx\,dy - \int_{\partial\Omega} k\frac{\partial u}{\partial n}v\,ds = 0 \qquad (4.3.7)$$

Substitution of the natural boundary condition in (4.2.18d), for example, into the boundary integral in (4.3.7) gives the variational equation

$$\int_\Omega (k\,\nabla u \cdot \nabla v + buv - fv)\,dx\,dy + \int_{\partial\Omega} p(u - \hat{u})v\,ds = 0 \qquad (4.3.8)$$

which must hold for all admissible test functions v.

Since we require that $u = \hat{u}$ on $\partial\Omega_1$,

$$\int_{\partial\Omega} p(u - \hat{u})v\, ds = \int_{\partial\Omega_2} puv\, ds - \int_{\partial\Omega_2} \gamma v\, ds$$

where we have denoted $\gamma = p\hat{u}$, Hence, our problem becomes one of finding a function u such that $u = \hat{u}$ on $\partial\Omega_1$ and

$$\int_{\Omega} (k\, \boldsymbol{\nabla} u \cdot \boldsymbol{\nabla} v + buv)\, dx\, dy + \int_{\partial\Omega_2} puv\, ds = \int_{\Omega} fv\, dx\, dy + \int_{\partial\Omega_2} \gamma v\, ds$$

$$(4.3.9)$$

for all admissible test functions v.

As in earlier discussions, we will regard (4.3.9) as the *given* boundary-value problem. Then, if (4.3.9) holds for all smooth test functions v and if the data and the solution u are sufficiently smooth, a solution of (4.3.9) will also be a solution of (4.2.18). Conversely, any solution of (4.2.18) will always automatically satisfy (4.3.9).

There remains the important issue of specifying the appropriate class of admissible functions for problem (4.3.9). We observe that the area integrals in (4.3.9) are well defined whenever u and v and their first partial derivatives are smooth enough to be square-integrable over Ω. Thus, we require that all admissible functions v be such that

$$\int_{\Omega} \left[\left(\frac{\partial v}{\partial x} \right)^2 + \left(\frac{\partial v}{\partial y} \right)^2 + v^2 \right] dx\, dy < \infty \qquad (4.3.10)$$

Adopting a minor modification in the notation used in previous chapters, we refer to the class of functions satisfying (4.3.10) as $H^1(\Omega)$, the superscript "1" reflecting the fact that first derivatives are square-integrable and (Ω) indicating the domain over which these functions are defined.

As was the case with one-dimensional problems, note that the natural boundary conditions (the conditions on $\partial\Omega_2$) enter (4.3.9) in the statement of the problem itself. These conditions appear in the terms $\int_{\partial\Omega_2} puv\, ds$ and $\int_{\partial\Omega_2} \gamma v\, ds$. The essential boundary conditions enter the problem in the definition of the classes of admissible functions. We choose as test functions those functions v in $H^1(\Omega)$ such that $v = 0$ on $\partial\Omega_1$. The solution u must be a function in $H^1(\Omega)$ such that $u = \hat{u}$ on $\partial\Omega_1$.

Our variational boundary-value problem can now be stated concisely in the following form:

> Find a function $u \in H^1(\Omega)$ such that $u = \hat{u}$
> on $\partial\Omega_1$ and (4.3.9) holds for all $v \in H^1(\Omega)$ (4.3.11)
> such that $v = 0$ on $\partial\Omega_1$.

In summary, we have obtained a variational formulation of the boundary-value problem that has the following attributes:

1. Any solution of (4.2.18) satisfies (4.3.9), and conversely, a sufficiently smooth solution of (4.3.9) is a solution of (4.2.18).

2. The variational formulation provides for statements and solutions of our physical boundary-value problems with weaker restrictions on the data than that required in (4.2.18).

3. The jump terms in (4.2.18b) do not require special attention in the variational statement.

4. The variational statement is the ideal starting point for the construction of approximate solutions using the finite element method.

EXERCISES

4.3.1 Expanding the first term of (4.3.1) in component form gives

$$-\nabla \cdot (k\,\nabla u) = -\frac{\partial}{\partial x}\left(k\,\frac{\partial u}{\partial x}\right) - \frac{\partial}{\partial y}\left(k\,\frac{\partial u}{\partial y}\right)$$

    (a) Derive the identity (4.3.2).

    (b) Expand (4.3.3) into component form and continue the development given in the text through (4.3.7) in this form.

4.3.2 Consider the boundary-value problem,

$$-\Delta u + \lambda u = f \quad \text{in } \Omega$$
$$u = 0 \quad \text{on } \partial\Omega$$

where λ is a constant.

    (a) Develop a variational statement of this problem using Green's formula from Exercise 4.2.4.

    (b) Suppose that Ω is a square with sides of length a parallel to the x, y-coordinate axes, the origin being located at a corner. Consider the function

$$u = A \sin \frac{\pi x}{a} \sin \frac{\pi y}{a}$$

        defined on Ω, where A is a constant. With this example, show that the constant λ cannot be selected arbitrarily whenever $f(x, y) \neq 0$ if our problem is to have a solution. Indeed, show that if λ is an *eigenvalue* of the Laplacian (i.e., if $\Delta u = \lambda u$, $u \neq 0$), then there does not exist a

u such that $-\Delta u + \lambda u = f$ for $f \not\equiv 0$. Discuss this situation for cases in which $\lambda \geq 0$ and $\lambda < 0$.

4.3.3   Suggest a modification of (4.3.9) to include the effect of a concentrated point source $\hat{f}\delta(x - x_0, y - y_0)$ at the point (x_0, y_0).

4.3.4   Consider the presence of a line source, that is, a source distribution that supplies the flux in a finite amount *per unit length* along a curve L in Ω, so that the jump in σ_n across L has a prescribed finite value, say f^L. Develop the appropriate jump condition and incorporate it into the variational equation (4.3.9).

4.3.5   Consider the special case in which $b \equiv 0$ and the condition

$$-k(s)\frac{\partial u(s)}{\partial n} = \hat{\sigma}(s)$$

is imposed on all of $\partial\Omega$ (i.e., $\partial\Omega_2 = \partial\Omega$) in (4.3.9).

(a)   Show that if u is a solution of the variational problem, then $u + c$ is also a solution, c being an arbitrary constant.

(b)   Show that compatibility condition (4.2.19) is a necessary condition for the existence of a solution to the variational problem.

[*HINT:*   Consider $u = c$, $v = 1$.]

4.4 FINITE ELEMENT INTERPOLATION

4.4.1 Discretization

This stage of our analysis represents a direct but nontrivial extension of the ideas discussed earlier for one-dimensional problems. Having a variational statement of our model problem, we proceed to construct a finite element mesh representing Ω. In the one-dimensional problem, this amounted to partitioning an interval into line elements connected at nodal points at their ends. For two-dimensional problems, the discretization of Ω is less straightforward.

The basic idea is to continue to represent approximate solutions u_h and test functions v_h by polynomials defined piecewise over geometrically simple subdomains of some region Ω_h, with Ω_h now in the x, y-plane. Our first concern is to choose a discretization that will be general enough to model irregular domains but consist of elements simple enough to minimize computational effort. As indicated in Fig. 4.7, simple triangles and/or quadrilaterals can be used for this purpose. If $\partial\Omega$ is curved, as in the figure, there will always be some *discretization error*, since the finite element mesh Ω_h, constructed as the collection of triangular or quadrilateral elements, will not

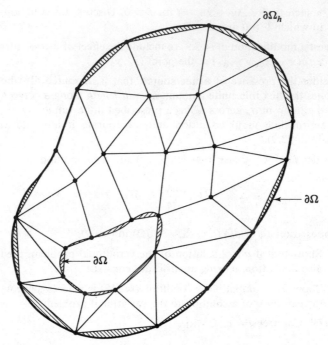

FIGURE 4.7   *A finite element discretization of a domain Ω by a mesh*
Ω_h consisting of triangular and quadrilateral elements.

perfectly coincide with the given domain Ω.* However, as the mesh is refined, Ω_h can approximate Ω with increasing accuracy.

Another reason for considering elements of simple shapes such as triangles is that there is a natural correspondence between the number and location of nodal points in an element and the number of terms used in the local polynomial approximation. Recall that for piecewise-linear approximations in one dimension, the restriction of a test function to an element Ω_e was a linear function of the form

$$v_h^e(x) = a_1 + a_2 x$$

a_1 and a_2 being constants. Since the element has two nodes, the constants a_1 and a_2 are uniquely determined by specifying the value of v_h^e at the endpoints of the element. Having done this, a continuous function v_h is produced by demanding that functions v_h^e and v_h^{e+1} in adjacent elements have the same value at their common node.

An analogous situation exists in two dimensions. A linear function in

---

* It is possible to use finite elements with curved boundaries that fit $\partial\Omega$ very closely or, at times, exactly. We give examples of such elements in Chapter 5.

two dimensions is of the form

$$v_h(x, y) = a_1 + a_2x + a_3y$$

with three constants: a_1, a_2, and a_3. Thus, three independent values of v_h must be specified to determine these constants, which means that the element should have three nodes, suggesting a triangle with nodes at the vertices. Moreover, if two adjacent triangles in the mesh share one side (and, hence, share two nodes), a function continuous across the interface of these elements will be produced by demanding that the linear functions on each element have the same values at the common nodes.

Similarly, a bilinear function

$$v_h(x, y) = a_1 + a_2x + a_3y + a_4xy$$

has four constants and might qualify as a shape function for an element with four nodes (a rectangle). Likewise the quadratic

$$v_h(x, y) = a_1 + a_2x + a_3y + a_4x^2 + a_5xy + u_6y^2$$

having six parameters, could be used to construct an element with six nodes (e.g., a triangle with a node at each vertex and at the midpoint of each side), and so on.

We now furnish some details as to how such two-dimensional finite element representations can be constructed. As in Chapter 2, it is informative to consider the finite element concept as a device for interpolating a given function $g = g(x, y)$ defined on Ω. As before, our aim is to construct an interpolant g_h of g of the form

$$g_h(x, y) = \sum_{j=1}^{N} g_j \phi_j(x, y), \qquad (x, y) \in \Omega_h \tag{4.4.1}$$

where $\phi_1(x, y)$, $\phi_2(x, y)$, ..., $\phi_N(x, y)$ are basis functions defined over Ω_h satisfying

$$\phi_i(x_j, y_j) = \begin{cases} 1 & \text{if } i = j \\ 0 & \text{if } i \neq j \end{cases} \tag{4.4.2}$$

where (x_j, y_j) are the coordinates of nodal points in the finite element mesh. When (4.4.2) holds, we have

$$g_h(x_j, y_j) = g_j, \qquad j = 1, 2, \ldots, N \tag{4.4.3}$$

so that by setting $g_j = g(x_j, y_j)$, g_h will coincide with (and, therefore, interpolate) the given function g at the nodes.

We must deal with two basic requirements:

1. The construction of local shape functions ψ_i^e defined over each element Ω_e in the mesh, must be such that when patched together in the manner indicated in previous chapters, they produce basis functions satisfying (4.4.2).

2. In anticipation of solving our model problem, the resulting basis functions ϕ_i must be square-integrable and have square-integrable first partial derivatives; that is, they must satisfy

$$\int_{\Omega_h} \left[\left(\frac{\partial \phi_i}{\partial x} \right)^2 + \left(\frac{\partial \phi_i}{\partial y} \right)^2 + \phi_i^2 \right] dx \, dy < \infty, \qquad i = 1, 2, \ldots, N$$

(4.4.4)

This requirement is satisfied by constructing the functions ϕ_i to be continuous across interelement boundaries.

4.4.2 Piecewise-Linear Interpolation on Triangles

Since the linear function,

$$v_h(x, y) = a_1 + a_2 x + a_3 y$$

determines a plane surface, the use of linear interpolation on a triangle will result in the approximation of a given smooth function v by a planar function of the type shown in Fig. 4.8. Suppose that Ω_h consists of a collection of E triangular elements and that we consider such a linear interpolation over a typical finite element Ω_e. Then, the restriction of v_h to Ω_e will be of the form*

$$v_h^e(x, y) = a_1 + a_2 x + a_3 y \qquad \text{for } (x, y) \in \Omega_e \qquad (4.4.5)$$

We determine the three constants from the conditions

$$v_1 = v_h^e(x_1, y_1) = a_1 + a_2 x_1 + a_3 y_1$$
$$v_2 = v_h^e(x_2, y_2) = a_1 + a_2 x_2 + a_3 y_2$$
$$v_3 = v_h^e(x_3, y_3) = a_1 + a_2 x_3 + a_3 y_3$$

---

* Our use of a local Cartesian coordinate system here for describing v_h^e is not always optimal for computational purposes. We will discuss more natural choices of coordinates for element calculations in Section 5.4.

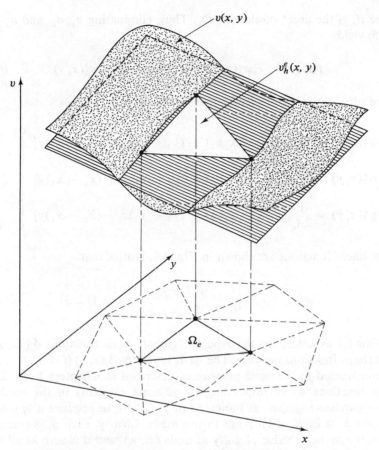

FIGURE 4.8 *Illustration of the idea that the three values of v_h^e at the vertices of a triangular element Ω_e determine a plane which intersects surface $v = v(x, y)$ at three points.*

where (x_i, y_i), $i = 1, 2, 3$, are the coordinates of the three vertices of the triangle. Solving this system for a_1, a_2, and a_3, we find

$$a_1 = \frac{1}{2A_e}[v_1(x_2 y_3 - x_3 y_2) + v_2(x_3 y_1 - x_1 y_3) + v_3(x_1 y_2 - x_2 y_1)]$$

$$a_2 = \frac{1}{2A_e}[v_1(y_2 - y_3) + v_2(y_3 - y_1) + v_3(y_1 - y_2)]$$

$$a_3 = \frac{1}{2A_e}[v_1(x_3 - x_2) + v_2(x_1 - x_3) + v_3(x_2 - x_1)]$$

where A_e is the area* of element Ω_e. Thus, eliminating a_1, a_2, and a_3 from (4.4.5) yields

$$v_h^e(x, y) = v_1\psi_1^e(x, y) + v_2\psi_2^e(x, y) + v_3\psi_3^e(x, y) \qquad (4.4.6)$$

where $\psi_i^e(x, y)$ are the element shape functions,

$$\left. \begin{aligned} \psi_1^e(x, y) &= \frac{1}{2A_e}[(x_2 y_3 - x_3 y_2) + (y_2 - y_3)x + (x_3 - x_2)y] \\ \psi_2^e(x, y) &= \frac{1}{2A_e}[(x_3 y_1 - x_1 y_3) + (y_3 - y_1)x + (x_1 - x_3)y] \\ \psi_3^e(x, y) &= \frac{1}{2A_e}[(x_1 y_2 - x_2 y_1) + (y_1 - y_2)x + (x_2 - x_1)y] \end{aligned} \right\} \qquad (4.4.7)$$

These linear functions are shown in Fig. 4.9. Notice that

$$\psi_i^e(x_j, y_j) = \begin{cases} 1 & \text{if } i = j, \\ 0 & \text{if } i \neq j \end{cases} \qquad i, j = 1, 2, 3 \qquad (4.4.8)$$

Now let us determine the type of "global" basis functions $\phi_i(x, y)$ that these shape functions produce. The basis functions $\phi_i(x, y)$ $(i = 1, 2, \ldots, N)$ are constructed in the same manner as described in Chapters 1 and 2: the shape functions ψ_i^e corresponding to adjacent elements in the mesh are simply patched together, as indicated in Fig. 4.10, to produce a "pyramid" function ϕ_i at each nodal point in the mesh. Clearly, each ϕ_i is piecewise-linear, it assumes a value of unity at node (x_i, y_i), and it is zero at all other nodes (x_j, y_j), $j \neq i$, and, therefore, satisfies (4.4.2). For a boundary node, we have an analogous situation in that ϕ_i assumes the form of a portion of a "pyramid." Of equal importance, the functions produced in this way are continuous across interelement boundaries and, therefore, over Ω_h; their first partial derivatives are step functions and, hence, are square-integrable. Thus, such basis functions would be appropriate choices for constructing

---

* Technically, the denominator here is

$$\begin{vmatrix} 1 & x_1 & y_1 \\ 1 & x_2 & y_2 \\ 1 & x_3 & y_3 \end{vmatrix} = x_2 y_3 + x_1 y_2 + x_3 y_1 - y_1 x_2 - y_2 x_3 - y_3 x_1$$

which equals twice the area A_e of the triangle whenever a right-handed coordinate system is used and the nodes are numbered consecutively according to the right-hand rule. If the same numbering is used in a left-handed system, this determinant equals $-2A_e$.

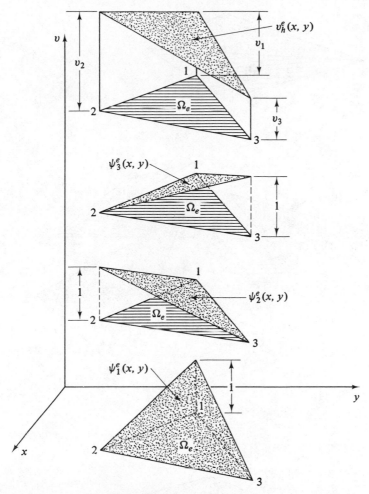

FIGURE 4.9 *Linear shape functions ψ_i^e over a triangular element Ω_e and their linear combination $v_h^e(x, y) = v_1\psi_1^e(x, y) + v_2\psi_2^e(x, y) + v_3\psi_3^e(x, y)$, $(x, y)\epsilon\Omega_e$.*

finite element approximations of the model problems discussed in the preceding section.

4.4.3   Other Triangular Elements

Other triangular elements involving higher-degree polynomials in x and y can be easily constructed. Let us first display the terms appearing in poly-

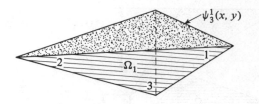

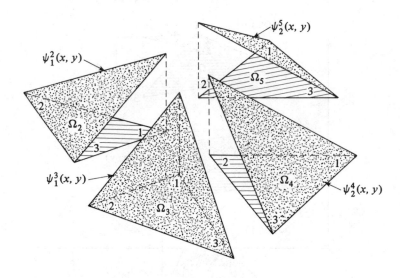

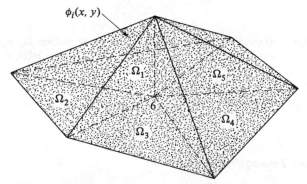

FIGURE 4.10 *Patching together linear shape functions ψ_i^e on five elements sharing i so as to produce a piecewise-linear basis function ϕ_i corresponding to node i.*

nomials of various degree in two variables in the following tabular form:

$$1 \qquad\qquad \text{degree 0}$$
$$x \ \ y \qquad\qquad \text{degree 1}$$
$$x^2 \ \ xy \ \ y^2 \qquad\qquad \text{degree 2}$$
$$x^3 \ \ x^2y \ \ xy^2 \ \ y^3 \qquad\qquad \text{degree 3}$$
$$x^4 \ \ x^3y \ \ x^2y^2 \ \ xy^3 \ \ y^4 \qquad\qquad \text{degree 4}$$
$$\vdots \qquad\qquad \vdots$$

This triangular array is called *Pascal's triangle*. Note that a complete poly-nomial of degree k in x and y will have exactly $\frac{1}{2}(k + 1)(k + 2)$ terms. Thus, a polynomial of degree k can be uniquely determined by specifying its value at $\frac{1}{2}(k + 1)(k + 2)$ points in the plane. Moreover, the location of entries in Pascal's triangle suggests a symmetric location of nodal points in triangular elements that will produce exactly the right number of nodes to define a polynomial interpolant of any degree. For instance, the six terms in a quadratic polynomial will be determined by specifying the value of v_h^e at six nodal points in a triangle, one at each vertex and one at the midpoint of each side—precisely the location of entries in the triangle formed by the quadratic in Pascal's triangle. Similarly, a complete cubic, having 10 terms, leads to a triangular element with 10 nodes. The location of the nodes is, again, deter-mined by Pascal's triangle: one at each vertex, two on each side dividing each side into three equal lengths, and one at the centroid. Similarly, a complete quartic leads to 15 nodes, and so on. The family of finite elements generated in this manner is illustrated in Fig. 4.11a.

Another important feature of these elements is that they produce, for polynomials of degree > 0, basis functions that are continuous over the domain and, therefore, have square-integrable first partial derivatives. Con-sider, for example, two adjacent six-node triangles Ω_e and Ω_{e+1} in the mesh. The local interpolants v_h^e and v_h^{e+1} are quadratic polynomials that must coincide at the three nodal points common to each element. However, the specification of three values of a quadratic in one dimension uniquely deter-mines that quadratic. Hence, v_h^e and v_h^{e+1} will coincide *everywhere* on the common boundary of the two elements, and v_h will, therefore, be continuous across this boundary. The idea is illustrated in Fig. 4.11b. Similarly, for cubic elements, values at four nodes are matched on common boundaries. Since a one-dimensional cubic is determined by specifying four independent values, a continuous piecewise cubic function is formed by patching together such elements.

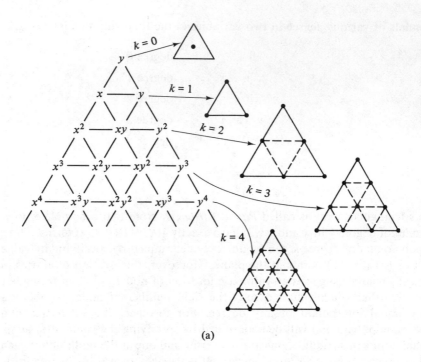

(a)

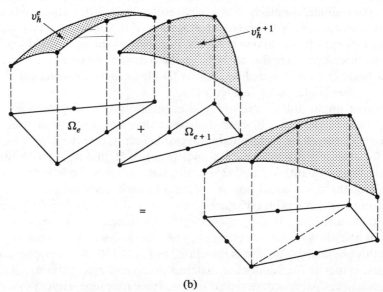

(b)

FIGURE 4.11 (a) Use of Pascal's triangle to generate various triangular elements over which complete polynomial shape functions of any degree k are defined; and (b) illustration, for the case k = 2, that basis functions produced by such elements are continuous across interelement boundaries.

158

4.4.4. Rectangular Elements

By taking the "product" of sets of polynomials in x with sets of polynomials in y, shape functions for a variety of rectangular elements can be obtained. In Section 5.4, we show how these ideas can be used to construct general quadrilateral elements. For example, a linear polynomial in x is characterized by a linear combination of the monomials $(1, x)$. The *tensor product* with monomials $(1, y)$ produces the matrix of four functions

$$\begin{bmatrix} 1 \\ x \end{bmatrix} [1, y] = \begin{bmatrix} 1 & y \\ x & xy \end{bmatrix} \tag{4.4.9}$$

and a linear combination of the entries in this matrix produces a local bilinear polynomial of the form

$$v_h^e(x, y) = a_1 + a_2 x + a_3 y + a_4 xy \tag{4.4.10}$$

There are four constants in (4.4.10) and four elements in the tensor product (4.4.9). Thus, if we visualize a rectangular element with four nodes, one at each corner, the function v_h^e in (4.4.10) is completely and uniquely determined by specifying its values at these four nodal points. Moreover, along the sides $x = $ constant, $y = $ constant, v_h^e is linear in x or y. Thus, if two such rectangular elements Ω_e and Ω_{e+1} have a common side in the mesh, a function that is continuous across their common interelement boundary will be produced by demanding that v_h^e and v_h^{e+1} assume the same values at nodes common to each element. Hence, shape functions obtained using (4.4.10) will produce basis functions ϕ_i which have square-integrable first derivatives over Ω_h.

By considering tensor products of polynomials of higher degree, element shape functions can be constructed which contain polynomials of any desired degree and which lead to basis functions that are continuous throughout Ω_h. For example, the product of two quadratics yields a matrix with nine entries:

$$\begin{bmatrix} 1 \\ x \\ x^2 \end{bmatrix} [1 \quad y \quad y^2] = \begin{bmatrix} 1 & y & y^2 \\ x & xy & xy^2 \\ x^2 & x^2y & x^2y^2 \end{bmatrix} \tag{4.4.11}$$

A biquadratic local interpolant v_h^e is obtained by forming a linear combination of all nine terms in this matrix. By constructing a rectangular element with nine nodes, one node located in the element at the point corresponding to the location of each entry in the foregoing matrix, an element is produced which leads to piecewise-biquadratic basis functions continuous on all of Ω_h. Similarly, a tensor product of cubics leads to an element with 16 nodes

159

and bicubic shape functions; and so on. Various rectangular elements produced by tensor products of polynomials are illustrated in Fig. 4.12.

4.4.5 Interpolation Error

Suppose that a smooth function g is given. Further, assume that we wish to interpolate g by a finite element representation g_h which contains complete polynomials of degree k. As in the one-dimensional case in (2.6.12), if partial

	1	y	y^2	y^3	y^4 $\cdots$
1	1	y	y^2	y^3	y^4 $\cdots$
x	x	xy	xy^2	xy^3	xy^4 $\cdots$
x^2	x^2	x^2y	x^2y^2	x^2y^3	x^2y^4 $\cdots$
x^3	x^3	x^3y	x^3y^2	x^3y^3	x^3y^4 $\cdots$
x^4	x^4	x^4y	x^4y^2	x^4y^3	x^4y^4 $\cdots$
$\vdots$	$\vdots$	$\vdots$	$\vdots$	$\vdots$	$\vdots$

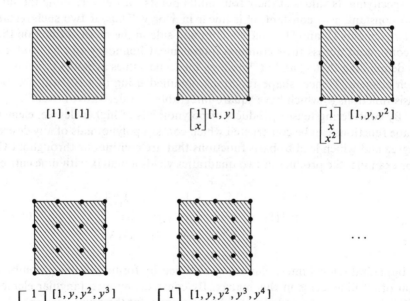

$[1] \cdot [1]$

$\begin{bmatrix} 1 \\ x \end{bmatrix}[1,y]$

$\begin{bmatrix} 1 \\ x \\ x^2 \end{bmatrix} [1,y,y^2]$

$\begin{bmatrix} 1 \\ x \\ x^2 \\ x^3 \end{bmatrix} [1,y,y^2,y^3]$

$\begin{bmatrix} 1 \\ x \\ x^2 \\ x^3 \\ x^4 \end{bmatrix} [1,y,y^2,y^3,y^4]$

$\cdots$

FIGURE 4.12 *Matrix containing terms of a tensor product of polynomials and various rectangular elements obtained using such tensor products.*

derivatives of g of order $k + 1$ are bounded in Ω_e, the interpolation error satisfies

$$\|g - g_h\|_{\infty, \Omega_e} = \max_{(x,y) \in \Omega_e} |g(x, y) - g_h(x, y)| \qquad (4.4.12)$$
$$\leq C h_e^{k+1}$$

where C is a positive constant and h_e is the "diameter" of Ω_e; that is, h_e is the largest distance between any two points in Ω_e. As in the one-dimensional case, note that this estimate holds only if all terms in a complete polynomial of degree k appear in g_h. Similarly,

$$\left\| \frac{\partial g}{\partial x} - \frac{\partial g_h}{\partial x} \right\|_{\infty, \Omega_e} \leq C_1 h_e^k \quad \text{and} \quad \left\| \frac{\partial g}{\partial y} - \frac{\partial g_h}{\partial y} \right\|_{\infty, \Omega_e} \leq C_2 h_e^k \quad (4.4.13)$$

The H^1-norm in two dimensions is defined by

$$\|g\|_1^2 = \int_\Omega \left[g^2 + \left(\frac{\partial g}{\partial x} \right)^2 + \left(\frac{\partial g}{\partial y} \right)^2 \right] dx \, dy \qquad (4.4.14)$$

Suppose that $\Omega_h = \Omega$ and h is the maximum diameter of all elements in the mesh. It can also be shown that for reasonable meshes and refinements,

$$\|g - g_h\|_1 < C_3 h^k \qquad (4.4.15)$$

for h sufficiently small. This estimate holds only if g_h contains a complete polynomial of degree k. For example, if g_h is piecewise-linear on triangles, $k = 1$ and we say that the H^1-interpolation error is $O(h)$. Similarly, if g_h is piecewise-bilinear, so that $g_h^e = a_1 + a_2 x + a_3 y + a_4 xy$, then (4.4.15) holds with $k = 1$ even though g_h^e contains a quadratic term xy. The key is that g_h^e contains a *complete* polynomial of degree $k = 1$ but not $k = 2$: the terms x^2 and y^2 are missing. Similarly, if g_h^e is a tensor product of quadratics, it will only contain complete polynomials of degree $k = 2$, even though cubic and quartic terms appear in such shape functions. These extra terms furnish enough nodal points for the element to provide for the generation of continuous basis functions, but they do not contribute to the asymptotic rate of convergence of the interpolation error.

EXERCISES

4.4.1 Consider a rectangular element with four nodes, one located at each corner of the element. In view of our criteria for acceptable finite-element basis functions, why is the following choice of a local test function representation unacceptable?

$$v_h^e(x, y) = a_1 + a_2 x + a_3 y + a_4 x^2$$

Here $a_1, a_2, a_3,$ and a_4 are constants.

4.4.2 Derive explicit formulas for the shape functions $\psi_i^e(x, y)$, $i = 1, 2, 3, 4$, for piecewise-bilinear interpolation on a rectangular element. Sketch the shape function corresponding to one of the four nodal points. Also sketch typical global basis functions ϕ_i generated on a patch of four such elements.

4.4.3 Suppose that Ω_h is a square consisting of eight triangular elements of equal size. Describe by means of sketches, the global basis functions ϕ_i, $i = 1, 2,$ $\ldots, 9$, generated by piecewise-linear shape functions on each element.

4.5 FINITE ELEMENT APPROXIMATIONS

4.5.1 Approximation of Two-Dimensional Boundary-Value Problems

Let us now return to the problem described in Sections 4.2 and 4.3. In particular, consider the general variational boundary-value problem (4.3.11). Let $H^1(\Omega)$ denote the class of functions satisfying (4.3.10) and defined over the whole domain Ω. Our problem is then to find a function u in $H^1(\Omega)$ such that $u = \hat{u}$ on $\partial\Omega_1$ and such that

$$\int_\Omega \left[k\left(\frac{\partial u}{\partial x}\frac{\partial v}{\partial x} + \frac{\partial u}{\partial y}\frac{\partial v}{\partial y}\right) + buv \right] dx\, dy + \int_{\partial\Omega_2} puv\, ds$$
$$= \int_\Omega fv\, dx\, dy + \int_{\partial\Omega_2} \gamma v\, ds \tag{4.5.1}$$

for all $v \in H^1(\Omega)$ such that $v = 0$ on $\partial\Omega_1$ and where $\gamma = p\hat{u}$.

The approximation of (4.5.1) follows the pattern now familiar from our discussions in Chapters 1 and 2. We replace Ω by a domain Ω_h that consists of a collection of E finite elements and N nodal points and we define an N-dimensional subspace H^h of $H^1(\Omega_h)$ by constructing an appropriate set of global basis functions ϕ_i, $i = 1, 2, \ldots, N$, using elements of the type described in the preceding section. Since the shape functions are continuous within each element, they cannot model a jump in material properties there. Hence, as in Chapter 2, we always choose the location of nodes and element boundaries to coincide (as closely as possible) with interfaces at which jumps in the modulus k occur, as indicated in Fig. 4.13. A typical test function in H^h will be of the form

$$v_h(x, y) = \sum_{j=1}^N v_j\phi_j(x, y) \tag{4.5.2}$$

where $v_j = v_h(x_j, y_j)$. In general, the Dirichlet data $\hat{u}$ given on $\partial\Omega_1$ is approximated by its interpolant $\hat{u}_h(s) = \sum \hat{u}_j\phi_j(x(s), y(s))$, the sum being taken over all nodes on the approximation $\partial\Omega_{1h}$ of $\partial\Omega_1$. Our approximation of (4.5.1)

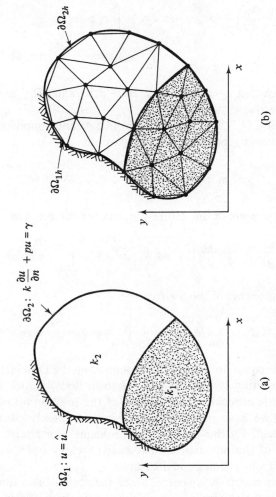

$$\partial\Omega_2: k\frac{\partial u}{\partial n} + pu = \gamma$$

$$\partial\Omega_1: u = \hat{u}$$

(a)

(b)

FIGURE 4.13 (a) The domain Ω for the model problem and (b) its finite-element discretization Ω_h.

163

then consists of seeking a function u_h in H^h,

$$u_h(x, y) = \sum_{j=1}^{N} u_j \phi_j(x, y) \tag{4.5.3}$$

such that $u_j = \hat{u}_j$ at the nodes on $\partial\Omega_{1h}$ and

$$\int_{\Omega_h} \left[k\left(\frac{\partial u_h}{\partial x}\frac{\partial v_h}{\partial x} + \frac{\partial u_h}{\partial y}\frac{\partial v_h}{\partial y}\right) + b u_h v_h \right] dx\, dy + \int_{\partial\Omega_{2h}} p u_h v_h\, ds$$
$$= \int_{\Omega_h} f v_h\, dx\, dy + \int_{\partial\Omega_{2h}} \gamma v_h\, ds \tag{4.5.4}$$

for all $v_h \in H^h$ such that $v_h = 0$ on $\partial\Omega_{1h}$. Here $\partial\Omega_{1h}$ and $\partial\Omega_{2h}$ approximate $\partial\Omega_1$ and $\partial\Omega_2$, respectively.

Upon substituting (4.5.2) and (4.5.3) into (4.5.4) and simplifying terms, we arrive at the linear algebraic system of equations

$$\sum_{j=1}^{N} K_{ij} u_j = F_i, \qquad i = 1, 2, \ldots, N \tag{4.5.5}$$

where K_{ij} are the elements of the stiffness matrix for the problem,

$$K_{ij} = \int_{\Omega_h} \left[k\left(\frac{\partial \phi_i}{\partial x}\frac{\partial \phi_j}{\partial x} + \frac{\partial \phi_i}{\partial y}\frac{\partial \phi_j}{\partial y}\right) + b\phi_i\phi_j \right] dx\, dy + \int_{\partial\Omega_{2h}} p\phi_i\phi_j\, ds \tag{4.5.6}$$

and F_i are the components of the load vector,

$$F_i = \int_{\Omega_h} f\phi_i\, dx\, dy + \int_{\partial\Omega_{2h}} \gamma\phi_i\, ds \tag{4.5.7}$$

We next modify the equations (4.5.5) to accommodate the Dirichlet data and finally solve the resulting system for the unknown nodal values u'_j thereby determining the finite element approximation of the solution u to (4.5.1).

The procedure we have just outlined, of course, closely parallels that described in Chapter 2 for one-dimensional problems. Fortunately, most of the other features of the one-dimensional analysis carry over, with minor modifications, to this two-dimensional case:

1. The stiffness matrix **K** is sparse. Since the global basis functions ϕ_i and ϕ_j and their derivatives are nonzero only on "patches" of elements containing nodes i and j, the entry K_{ij} will be nonzero only when there is an element containing both node i and node j.

2. In the present case, **K** is symmetric (owing to the fact that the operator in (4.2.18) is self-adjoint). Note also that if a judicious numbering of nodes is used, **K** will be *banded*; that is, the nonzero elements in **K** will form a band

containing the main diagonal of the matrix. The fact that $\mathbf{K}$ is a sparse, symmetric matrix can be thoroughly exploited in designing efficient algorithms* for solving linear systems of the form (4.5.5).

3.   Each of the integrals in (4.5.6) and (4.5.7) can be calculated as the sum of contributions furnished by each element in the mesh. However, the interpretation of such a procedure is interesting and deserves some elaboration. Let Ω_e denote a typical finite element in the mesh. On Ω_e, the exact solution u of our boundary-value problem satisfies

$$\int_{\Omega_e} (k \, \nabla u \cdot \nabla v + buv) \, dx \, dy = \int_{\Omega_e} fv \, dx \, dy - \int_{\partial\Omega_e} \sigma_n v \, ds$$

for every admissible v, where σ_n is the normal component of flux at the element boundary. Next, let u_h^e and v_h^e denote the restrictions of the approximations u_h and v_h to Ω_e. Then the local approximation of the variational boundary-value problem over Ω_e assumes the form

$$\int_{\Omega_e} (k \, \nabla u_h^e \cdot \nabla v_h^e + bu_h^e v_h^e) \, dx \, dy - \int_{\Omega_e} fv_h^e \, dx \, dy - \int_{\partial\Omega_e} \sigma_n v_h^e \, ds \qquad (4.5.8)$$

Here σ_n is the actual (exact) flux across $\partial\Omega_e$ and, although not given as data in the original problem, appears as data in natural boundary conditions on $\partial\Omega_e$. Since $v_h = 0$ on $\partial\Omega_{1h}$, there will be no contribution to the last integral of (4.5.8) from elements with sides coincident with $\partial\Omega_{1h}$.

Since u_h^e and v_h^e are of the form

$$u_h^e(x, y) = \sum_{j=1}^{N_e} u_j^e \psi_j^e(x, y), \qquad v_h^e(x, y) = \sum_{j=1}^{N_e} v_j^e \psi_j^e(x, y)$$

ψ_j^e being the local shape functions for Ω_e and N_e the number of nodes in Ω_e, (4.5.8) leads to the linear system

$$\sum_{j=1}^{N_e} k_{ij}^e u_j^e = f_i^e - \sigma_i^e, \qquad i = 1, 2, \ldots, N_e \qquad (4.5.9)$$

where

$$k_{ij}^e = \int_{\Omega_e} \left[k \left(\frac{\partial \psi_i^e}{\partial x} \frac{\partial \psi_j^e}{\partial x} + \frac{\partial \psi_i^e}{\partial y} \frac{\partial \psi_j^e}{\partial y} \right) + b \psi_i^e \psi_j^e \right] dx \, dy \qquad (4.5.10)$$

$$f_i^e = \int_{\Omega_e} f \psi_i^e \, dx \, dy \qquad (4.5.11)$$

$$\sigma_i^e = \int_{\partial\Omega_e} \sigma_n \psi_i^e \, ds \qquad (4.5.12)$$

---

* We discuss several such algorithms in Volume III.

Here k_{ij}^e and f_i^e are the components of the element stiffness matrix and load vector, respectively, for element Ω_e and $\boldsymbol{\sigma}_e$ is an *element flux vector*, obtained by assigning to node i of Ω_e a weighted average $\int_{\partial\Omega_e} \sigma_n \psi_i^e \, ds$ of the actual flux σ_n across $\partial\Omega_e$.

Formally, the global system of equations (4.5.5) is obtained by summing (4.5.9) over all E elements in the mesh. We expand the element matrices in (4.5.10), (4.5.11), and (4.5.12) to $N \times N$ and $(N \times 1)$-order matrices corresponding to the order of the global matrices in (4.5.6) and (4.5.7). For example, $\mathbf{k}^e$ will become an $N \times N$ matrix $\mathbf{K}^e$ with zeros everywhere except those rows and columns corresponding to nodes within element Ω_e and $\mathbf{f}^e$ and $\boldsymbol{\sigma}^e$ will be expanded to $N \times 1$ vectors $\mathbf{F}^e$ and $\boldsymbol{\Sigma}^e$ with nonzero entries only in those rows corresponding to nodes in Ω_e. Then the first terms in the global matrices in (4.5.6) and (4.5.7) are obtained as the sums

$$
\left.
\begin{aligned}
\sum_{e=1}^{E} \int_{\Omega_e} \left[k\left(\frac{\partial\phi_i}{\partial x}\frac{\partial\phi_j}{\partial x} + \frac{\partial\phi_i}{\partial y}\frac{\partial\phi_j}{\partial y}\right) + b\phi_i\phi_j \right] dx\, dy &= \sum_{e=1}^{E} K_{ij}^e \\
\sum_{e=1}^{E} \int_{\Omega_e} f\phi_i \, dx\, dy &= \sum_{e=1}^{E} F_i^e, \qquad i,j = 1,2,\ldots,N
\end{aligned}
\right\}
\tag{4.5.13}
$$

and

$$
\sum_{e=1}^{E} (K_{ij}^e u_j - F_i^e + \Sigma_i^e) = 0, \qquad i = 1,2,\ldots,N
\tag{4.5.14}
$$

Notice that the contributions to K_{ij} and F_i from boundary conditions (recall (4.5.6) and (4.5.7)) must enter the problem through the terms Σ_i^e. Continuing, we note that the sum of the contour integrals can be written in the form

$$
\sum_{e=1}^{E} \Sigma_i^e = S_i^{(0)} + S_i^{(1)} + S_i^{(2)}, \qquad i = 1,2,\ldots,N
\tag{4.5.15}
$$

where

$$
\begin{aligned}
S_i^{(0)} &= \sum_{e=1}^{E} \int_{\partial\Omega_e - \partial\Omega_h} \sigma_n \phi_i \, ds \\
S_i^{(1)} &= \int_{\partial\Omega_{1h}} \sigma_n \phi_i \, ds \\
S_i^{(2)} &= \int_{\partial\Omega_{2h}} \sigma_n \phi_i \, ds
\end{aligned}
\tag{4.5.16}
$$

Here $\partial\Omega_e - \partial\Omega_h$ is the portion of the boundary $\partial\Omega_e$ of Ω_e not on $\partial\Omega_h$ (i.e., that part of $\partial\Omega_e$ that consists of interelement boundaries). We interpret the quantities in (4.5.16) as follows.

$S_i^{(0)}$: Since $S_i^{(0)}$ only involves terms on $\partial\Omega_e - \partial\Omega_h$, this vector is defined only at interior nodes i. To interpret $S_i^{(0)}$, consider an interior patch of four elements having node 1 in common such as that indicated in Fig. 4.14. Using (4.5.8) and (4.5.14), we easily verify that, for this node, $S_1^{(0)}$ has the form

$$
S_1^{(0)} = \sum_{e=1}^{4} \int_{\partial\Omega_e} \sigma_n \phi_1 \, ds
$$

$$
= \int_{\Gamma_1} [\![\sigma_n]\!]\phi_1 \, ds + \int_{\Gamma_2} [\![\sigma_n]\!]\phi_1 \, ds + \int_{\Gamma_3} [\![\sigma_n]\!]\phi_1 \, ds + \int_{\Gamma_4} [\![\sigma_n]\!]\phi_1 \, ds
$$

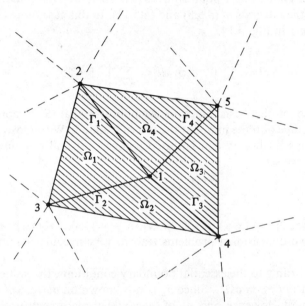

FIGURE 4.14   *An interior patch of four elements sharing node 1.*

From the conservation law, $[\![\sigma_n]\!] = 0$ across an interface where no point or line sources are applied. Thus, if f is smooth in the patch shown in Fig. 4.14, we have

$$
S_1^{(0)} = 0 \tag{4.5.17}
$$

Naturally, there is no need to evaluate these zero contributions to (4.5.14) and (4.5.15), so that they may be excluded from element calculations.

An exception to (4.5.17) occurs when the source function f contains a line source or concentrated point source. Then $[\![\sigma_n]\!]$ equals the intensity of the line source and is no longer zero. In the case of a point source, our present variational formulation is not strictly applicable. However, we can include the effects of point sources in the finite element analysis if we proceed as follows. Suppose that f has the form

$$f(x, y) = \bar{f}(x, y) + \hat{f}\delta(x - x_i, y - y_i) \tag{4.5.18}$$

where $\bar{f}$ is the smooth (integrable) part of f and $\hat{f}\delta(x - x_i, y - y_i)$ denotes a point source of intensity $\hat{f}$ at point $(x_i, y_i) \in \Omega_h$. As in our study of one-dimensional problems, the mesh Ω_h is always constructed so that nodal points are located at points where point sources act. Then only the smooth part $\bar{f}$ of f appears in the integrals in (4.5.8) and (4.5.11). In this case, note that for the interior node 1 in Fig. 4.14,

$$S_1^{(0)} = \sum_{e=1}^{4} \int_{\partial\Omega_e} \sigma_n \phi_1 \, ds = \sum_{m=1}^{4} \int_{\Gamma_m} [\![\sigma_n]\!]\phi_1 \, ds$$

The presence of the basis function ϕ_1 indicates that $S_1^{(0)}$ represents the weighted average of these jumps at the interior node 1. We choose to balance nonzero jumps in flux by concentrated sources $\hat{f}$ (recall our discussion in Section 2.2), we set

$$S_1^{(0)} = \hat{f}$$

whenever $f(x, y) = \bar{f}(x, y) + \hat{f}\delta(x - x_1, y - y_1)$. We remark that point sources in two-dimensional problems lead to very irregular (singular) solutions u.

$S_i^{(1)}$: According to the essential boundary conditions, the values of u_h are prescribed at nodes on $\partial\Omega_{1h}$. Since σ_n is not known on $\partial\Omega_{1h}$, $S_i^{(1)}$ cannot be prescribed there. However, once all of the nodal displacements $u_1, u_2, \ldots, u_N$ have been determined, an approximation of $S_i^{(1)}$ can be calculated directly from (4.5.14), if desired.

$S_i^{(2)}$: On $\partial\Omega_{2h}$, the natural boundary condition is specified. There we set

$$\sigma_n(s) = p(s)u_h(s) - \gamma(s)$$

so that, approximately,

$$S_i^{(2)} \approx \int_{\partial\Omega_{2h}} \left[p \sum_{j=1}^{N} u_j \phi_j - \gamma \right] \phi_i \, ds$$
$$= \sum_{j=1}^{N} P_{ij}u_j - \gamma_i \tag{4.5.19}$$

where

$$\gamma_i = \int_{\partial\Omega_{2h}} \gamma\phi_i \, ds = \sum_{e=1}^{E} \int_{\partial\Omega_{2h}^e} \gamma\phi_i \, ds = \sum_{e=1}^{E} \gamma_i^e \qquad (4.5.20)$$

and

$$P_{ij} = \int_{\partial\Omega_{2h}} p\phi_i\phi_j \, ds = \sum_{e=1}^{E} \int_{\partial\Omega_{2h}^e} p\phi_i\phi_j \, ds = \sum_{e=1}^{E} P_{ij}^e \qquad (4.5.21)$$

Here $\partial\Omega_{2h}^e$ is the portion of $\partial\Omega_e$ intersecting $\partial\Omega_{2h}$.

Returning now to (4.5.14), we arrive at the system of equations,

$$\sum_{j=1}^{N} K_{ij}u_j = F_i - S_i^{(1)}, \qquad i = 1, 2, \dots, N \qquad (4.5.22)$$

where

$$K_{ij} = \sum_{e=1}^{E} (K_{ij}^e + P_{ij}^e) \qquad (4.5.23)$$

$$F_i = \sum_{e=1}^{E} (F_i^e + \gamma_i^e) \qquad (4.5.24)$$

We now impose boundary conditions on $\partial\Omega_{1h}$ and proceed to solve the resulting system of equations for the unknown nodal values.

4.5.2 An Example

We shall briefly outline the analysis of a simple example problem, leaving most of the details as exercises. Consider the formulation of a finite element approximation of the problem

$$\left.\begin{array}{ll} -\Delta u(x, y) = f(x, y) & \text{in } \Omega \\[4pt] u = 0 & \text{on } \Gamma_{41} \\[4pt] \dfrac{\partial u}{\partial n} = 0 & \text{on } \Gamma_{12}, \Gamma_{25}, \Gamma_{67}, \text{ and } \Gamma_{74} \\[4pt] \dfrac{\partial u}{\partial n} + \beta u = \gamma & \text{on } \Gamma_{56} \end{array}\right\} \qquad (4.5.25)$$

where Ω is the polygonal domain shown in Fig. 4.15a and $\Gamma_{41}, \Gamma_{12}, \dots, \Gamma_{74}$ are the segments of the boundary indicated in the figure. In this case, $\partial\Omega_1 = \Gamma_{41}$ and $\partial\Omega_2$ consists of the segments $\Gamma_{12}, \Gamma_{25}, \Gamma_{56}, \Gamma_{67}$, and Γ_{74}.

Our analysis of this problem proceeds as follows:

1. We partition the domain into six triangular elements, as indicated in Fig. 4.15b, over which linear approximations u_h of the solution u of

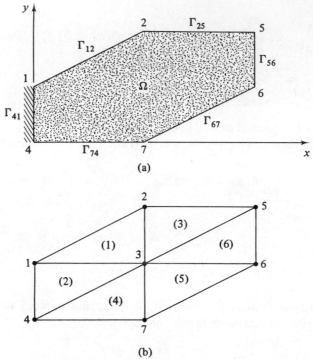

FIGURE 4.15 (a) The polygonal domain Ω in problem (4.5.25) and (b) a finite-element model of this domain.

(4.5.25) are defined. The six elements and seven nodes are numbered as shown. Note that $\partial\Omega_1 = \partial\Omega_{1h}$ and $\partial\Omega_2 = \partial\Omega_{2h}$.

2. Next, we use (4.5.10) and (4.5.11) to compute the element matrices $\mathbf{k}^e$ and $\mathbf{f}^e$, $e = 1, 2, \ldots, 6$, and expand these to 7×7 matrices:

Element 1

$$
\mathbf{K}^1 =
\begin{bmatrix}
k^1_{11} & k^1_{12} & k^1_{13} & 0 & 0 & 0 & 0 \\
k^1_{21} & k^1_{22} & k^1_{23} & 0 & 0 & 0 & 0 \\
k^1_{31} & k^1_{32} & k^1_{33} & 0 & 0 & 0 & 0 \\
0 & 0 & 0 & 0 & 0 & 0 & 0 \\
0 & 0 & 0 & 0 & 0 & 0 & 0 \\
0 & 0 & 0 & 0 & 0 & 0 & 0 \\
0 & 0 & 0 & 0 & 0 & 0 & 0
\end{bmatrix},
\quad
\mathbf{F}^1 =
\begin{bmatrix}
f^1_1 \\
f^1_2 \\
f^1_3 \\
0 \\
0 \\
0 \\
0
\end{bmatrix}
$$

Element 2

$$\mathbf{K}^2 = \begin{bmatrix} k_{11}^2 & 0 & k_{12}^2 & k_{13}^2 & 0 & 0 & 0 \\ 0 & 0 & 0 & 0 & 0 & 0 & 0 \\ k_{21}^2 & 0 & k_{22}^2 & k_{23}^2 & 0 & 0 & 0 \\ k_{31}^2 & 0 & k_{32}^2 & k_{33}^2 & 0 & 0 & 0 \\ 0 & 0 & 0 & 0 & 0 & 0 & 0 \\ 0 & 0 & 0 & 0 & 0 & 0 & 0 \\ 0 & 0 & 0 & 0 & 0 & 0 & 0 \end{bmatrix}, \qquad \mathbf{F}^2 = \begin{bmatrix} f_1^2 \\ 0 \\ f_2^2 \\ f_3^2 \\ 0 \\ 0 \\ 0 \end{bmatrix}$$

$$\vdots$$

Element 6

$$\mathbf{K}^6 = \begin{bmatrix} 0 & 0 & 0 & 0 & 0 & 0 & 0 \\ 0 & 0 & 0 & 0 & 0 & 0 & 0 \\ 0 & 0 & k_{11}^6 & 0 & k_{12}^6 & k_{13}^6 & 0 \\ 0 & 0 & 0 & 0 & 0 & 0 & 0 \\ 0 & 0 & k_{21}^6 & 0 & k_{22}^6 & k_{23}^6 & 0 \\ 0 & 0 & k_{31}^6 & 0 & k_{32}^6 & k_{33}^6 & 0 \\ 0 & 0 & 0 & 0 & 0 & 0 & 0 \end{bmatrix}, \qquad \mathbf{F}^6 = \begin{bmatrix} 0 \\ 0 \\ f_1^6 \\ 0 \\ f_2^6 \\ f_3^6 \\ 0 \end{bmatrix}$$

3. As the contributions from each element are calculated, starting from element 1 and continuing through element 6, they are added to their appropriate locations in the global stiffness and load matrices as indicated in (4.5.14). At this stage, we have the system

$$\begin{bmatrix} K_{11} & K_{12} & K_{13} & K_{14} & 0 & 0 & 0 \\ K_{21} & K_{22} & K_{23} & 0 & K_{25} & 0 & 0 \\ K_{31} & K_{32} & K_{33} & K_{34} & K_{35} & K_{36} & K_{37} \\ K_{41} & 0 & K_{43} & K_{44} & 0 & 0 & K_{47} \\ 0 & K_{52} & K_{53} & 0 & \tilde{K}_{55} & \tilde{K}_{56} & 0 \\ 0 & 0 & K_{63} & 0 & \tilde{K}_{65} & \tilde{K}_{66} & K_{67} \\ 0 & 0 & K_{73} & K_{74} & 0 & K_{76} & K_{77} \end{bmatrix} \begin{bmatrix} u_1 \\ u_2 \\ u_3 \\ u_4 \\ u_5 \\ u_6 \\ u_7 \end{bmatrix} = \begin{bmatrix} F_1 \\ F_2 \\ F_3 \\ F_4 \\ \tilde{F}_5 \\ \tilde{F}_6 \\ F_7 \end{bmatrix} - \begin{bmatrix} \Sigma_1 \\ 0 \\ 0 \\ \Sigma_4 \\ \Sigma_5 \\ \Sigma_6 \\ 0 \end{bmatrix}$$

$$(4.5.26)$$

where $F_1 = f_1^1 + f_1^2$, $F_2 = f_2^1 + f_2^3$, etc., and the Σ_i are defined using (4.5.15). The entries marked with $\sim$ will be modified in the final system of equations upon the application of the natural boundary conditions on Γ_{56}.

4. Since nonhomogeneous conditions are applied only on the segment connecting nodes 5 and 6, the matrices γ and P defined in (4.5.20) and (4.5.21) are of the form

$$\gamma = \begin{bmatrix} 0 \\ 0 \\ 0 \\ 0 \\ \gamma_5 \\ \gamma_6 \\ 0 \end{bmatrix}, \quad P = \begin{bmatrix} 0 & 0 & 0 & 0 & 0 & 0 & 0 \\ 0 & 0 & 0 & 0 & 0 & 0 & 0 \\ 0 & 0 & 0 & 0 & 0 & 0 & 0 \\ 0 & 0 & 0 & 0 & 0 & 0 & 0 \\ 0 & 0 & 0 & 0 & P_{55} & P_{56} & 0 \\ 0 & 0 & 0 & 0 & P_{65} & P_{66} & 0 \\ 0 & 0 & 0 & 0 & 0 & 0 & 0 \end{bmatrix} \quad (4.5.27)$$

Thus, (4.5.26) becomes

$$\begin{bmatrix} K_{11} & K_{12} & K_{13} & K_{14} & 0 & 0 & 0 \\ K_{21} & K_{22} & K_{23} & 0 & K_{25} & 0 & 0 \\ K_{31} & K_{32} & K_{33} & K_{34} & K_{35} & K_{36} & K_{37} \\ K_{41} & 0 & K_{43} & K_{44} & 0 & 0 & K_{47} \\ 0 & K_{52} & K_{53} & 0 & K_{55} & K_{56} & 0 \\ 0 & 0 & K_{63} & 0 & K_{65} & K_{66} & K_{67} \\ 0 & 0 & K_{73} & K_{74} & 0 & K_{76} & K_{77} \end{bmatrix} \begin{bmatrix} u_1 \\ u_2 \\ u_3 \\ u_4 \\ u_5 \\ u_6 \\ u_7 \end{bmatrix} = \begin{bmatrix} F_1 - \Sigma_1 \\ F_2 \\ F_3 \\ F_4 - \Sigma_4 \\ F_5 \\ F_6 \\ F_7 \end{bmatrix}$$

$$(4.5.28)$$

wherein

$$K_{55} = \tilde{K}_{55} + P_{55}, \qquad K_{56} = \tilde{K}_{56} + P_{56}, \qquad F_5 = \tilde{F}_5 + \gamma_5, \qquad \text{etc.}$$

5. We now impose the essential conditions $u_1 = u_4 = 0$ on Γ_{41}. In this way, we obtain the invertible system of five equations and five unknowns,

$$\begin{bmatrix} K_{22} & K_{23} & K_{25} & 0 & 0 \\ K_{32} & K_{33} & K_{35} & K_{36} & K_{37} \\ K_{52} & K_{53} & K_{55} & K_{56} & 0 \\ 0 & K_{63} & K_{65} & K_{66} & K_{67} \\ 0 & K_{73} & 0 & K_{76} & K_{77} \end{bmatrix} \begin{bmatrix} u_2 \\ u_3 \\ u_5 \\ u_6 \\ u_7 \end{bmatrix} = \begin{bmatrix} F_2 \\ F_3 \\ F_5 \\ F_6 \\ F_7 \end{bmatrix} \quad (4.5.29)$$

which we solve for the nodal values u_2, u_3, u_5, u_6, and u_7. The remaining pair of equations can then be used to calculate the approximate fluxes Σ_1 and Σ_4 at 1 and 4:

$$\left.\begin{aligned} -\Sigma_1 &= K_{12}u_2 + K_{13}u_3 + K_{14}u_4 - F_1 \\ -\Sigma_4 &= K_{43}u_3 + K_{47}u_7 - F_4 \end{aligned}\right\} \qquad (4.5.30)$$

Other features of the solution can now be evaluated since, by (4.5.3), u_h is now completely determined.

EXERCISE

4.5.1 Furnish additional details for the example problem in Section 4.5.2, as follows:

(a) Suppose that all of the elements in the mesh shown in Fig. 4.15a are equal isosceles triangles, the two equal sides being of length h. Derive the element stiffness matrix $\mathbf{k}^e$ and the load vector $\mathbf{f}^e$ for the element shown for the case $f(x, y) = 1$.

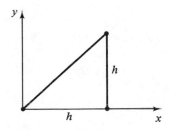

(b) Suppose that the coordinates of the nodes in the mesh shown in Fig. 4.15a are

$$(x_1, y_1) = (0, 1), \qquad (x_2, y_2) = (1, 2), \qquad (x_3, y_3) = (1, 1),$$
$$(x_4, y_4) = (0, 0), \qquad (x_5, y_5) = (2, 2), \qquad (x_6, y_6) = (2, 1),$$
$$(x_7, y_7) = (1, 0)$$

Using the results of part (a), calculate the element stiffness matrices and load vectors for all six elements, giving explicit numerical values for all entries.

(c) Sum the matrices derived in part (b) to obtain the matrices $\tilde{\mathbf{K}}$ and $\tilde{\mathbf{F}}$ of (4.5.26).

(d) Taking $p(s) = \gamma(s) = 1$ on Γ_{56}, compute the matrices $\boldsymbol{\gamma}$ and $\mathbf{P}$ of (4.5.27).

(e)   Using the results of parts (c) and (d), compute the final systems of equations (4.5.29) and (4.5.30) for this problem.

(f)   Describe what modifications in this analysis would be necessary in the case that nonhomogeneous boundary conditions are given on Γ_{41}. In particular,

   (1)   Describe the formulation of the case in which $u_1 = 2$, $u_4 = 3$.

   (2) $u = y$ on Γ_{41}.

   [*HINT:*   Consider an interpolation of $\hat{u} = y$].

(g)   Describe the modifications necessary in the analysis of this problem for the case in which, instead of the boundary conditions in (4.5.25), we have the Neumann conditions,

$$\frac{\partial u}{\partial n}(s) = \gamma(s) \quad \text{on all of } \partial\Omega; \qquad \gamma(s) = \begin{cases} \gamma_0 & \text{on } \Gamma_{41} \\ 0 & \text{on } \Gamma_{12} \text{ and } \Gamma_{25} \\ \gamma_1 & \text{on } \Gamma_{56} \\ 0 & \text{on } \Gamma_{67} \text{ and } \Gamma_{74} \end{cases}$$

where γ_0 and γ_1 are constants. If $f(x, y) = 1$, what are necessary conditions on γ_0 and γ_1 in order that there exists a solution to this problem?

5

TWO-DIMENSIONAL ELEMENT

CALCULATIONS

5.1 INTRODUCTION

In Chapter 4, we considered the variational formulation and finite-element approximation of a class of elliptic boundary-value problems in two dimensions. We now examine the necessary element calculations in greater detail.

Our objectives here are twofold: first, to describe a general and systematic method for calculating element matrices; and second, to describe how these techniques can be incorporated into a representative finite-element code for two-dimensional problems. Our treatment of both of these subjects parallels that for the one-dimensional problems considered in Chapters 2 and 3. The setting for calculations is, again, a master element $\hat{\Omega}$ on which all pertinent computations are made. For two-dimensional problems, the master element emerges as an extremely powerful and useful basis for implementing the finite-element method. Once certain details on the coding of such element calculations have been described, we show that these fit naturally into the same framework as the one-dimensional program developed in Chapter 3. Then, with very minor modifications, we can obtain a finite-element program, CODE2, for solving a general class of two-dimensional problems.

5.2 ELEMENT TRANSFORMATIONS

The first step in applying the finite element method to a given boundary-value problem is the construction of the mesh Ω_h. In this section, we consider a number of basic ideas concerning finite element meshes that provide a basis for a systematic and general technique for calculating element matrices.

5.2.1 The Master Element

The key concept in our approach is the notion of a master element $\hat{\Omega}$ similar to that described in Chapter 2 for one dimension. It is clear that the calculation of element stiffness matrices and load vectors for a curvilinear element Ω_e, such as that indicated in Fig. 5.1, would be awkward if performed directly in terms of the x, y-coordinates shown. Moreover, the character of

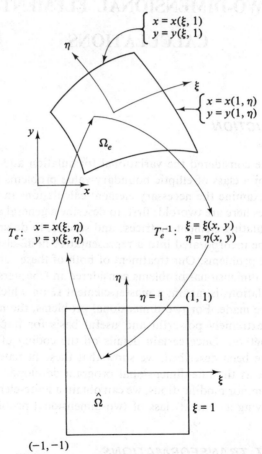

FIGURE 5.1 *A finite element Ω_e in the x, y-plane obtained as the image under T_e of the corresponding master element $\hat{\Omega}$ in the ξ, η-plane. The inverse map T_e^{-1} from Ω_e to $\hat{\Omega}$ is also indicated.*

such calculations (e.g., the limits of integration, etc.) would change from element to element in the mesh. If we introduce an invertible transformation between a master element $\hat{\Omega}$ of simple shape and an arbitrary element Ω_e, it should be possible to transform the operations on Ω_e so that they hold on $\hat{\Omega}$. Then we can perform the calculations conveniently on the master element. This "transformation" is accomplished by a simple coordinate transformation or mapping of points from $\hat{\Omega}$ into points in each element.

Consider, for instance, the master element $\hat{\Omega}$ shown in Fig. 5.1. To provide for the simplest possible calculations on $\hat{\Omega}$, we choose it to be a square, with local coordinates ξ and η scaled so that points (ξ, η) in $\hat{\Omega}$ satisfy $-1 \leq \xi \leq 1$, $-1 \leq \eta \leq 1$, as indicated. We then introduce a *map T_e* of $\hat{\Omega}$ onto Ω_e defined by the change of coordinates

$$T_e: \quad \left.\begin{array}{c} x = x(\xi, \eta) \\ y = y(\xi, \eta) \end{array}\right\} \qquad (5.2.1)$$

To interpret (5.2.1), consider, for instance, the line segment in $\hat{\Omega}$ consisting of points $(1, \eta)$. According to (5.2.1), these points are mapped into the points $x = x(1, \eta)$, $y = y(1, \eta)$ in the x, y-plane. But the points $((x(1, \eta), y(1, \eta))$ define parametrically a curve in the plane, with η now appearing as a real parameter. We say that this curve defines the *curvilinear coordinate line* $\xi = 1$ in the x, y-plane. Likewise, the line $\eta = 1$ in the master element is mapped into the curve $\eta = 1$ in the x, y-plane, defined by the parametric equations $x = x(\xi, 1)$, $y = y(\xi, 1)$. In fact, all vertical and horizontal lines $\xi = $ constant, $\eta = $ constant on $\hat{\Omega}$ are transformed into curvilinear coordinate lines $\xi = $ constant, $\eta = $ constant in the x, y-plane, as indicated in Fig. 5.1. We refer to element Ω_e as the *image* of $\hat{\Omega}$ under the map T_e defined by (5.2.1).

The basic idea of introducing the master element can now be stated: *the generation of the complete finite element mesh containing E elements is viewed as a sequence of transformations $\{T_1, T_2, \ldots, T_E\}$ of the form* (5.2.1) *in which each element Ω_e is the image of the fixed master element $\hat{\Omega}$ under a coordinate map T_e.** The process is illustrated symbolically in Fig. 5.2. We shall show that all of the intrinsic properties of a given type of finite element (number and location of nodes, shape functions, stiffness matrices, load vectors, etc.) can be prescribed for the fixed element $\hat{\Omega}$ and then carried to any element Ω_e in the mesh by using the map T_e.

There are several elementary properties of the coordinate transformation (5.2.1) that should be discussed before going on to the important computational considerations. First, let us suppose that the functions x and y are

---

* Of course, the possibility of more than a single master element being required in a given problem exists—for instance, when both triangular and quadrilateral elements are to be used. For simplicity, we shall restrict our attention to a single master element.

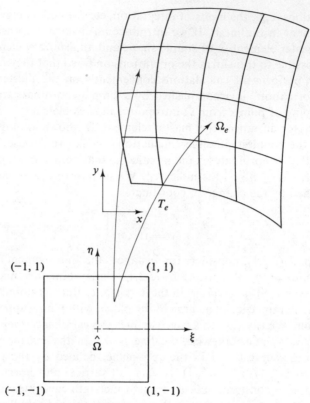

FIGURE 5.2 *Generation of a finite element mesh by invertible mappings from a master square.*

continuously differentiable with respect to ξ and η. Then the infinitesimals $d\xi$ and $d\eta$ transform into dx and dy according to

$$dx = \frac{\partial x}{\partial \xi} d\xi + \frac{\partial x}{\partial \eta} d\eta \quad \text{and} \quad dy = \frac{\partial y}{\partial \xi} d\xi + \frac{\partial y}{\partial \eta} d\eta$$

which can be written in matrix form as

$$\begin{bmatrix} dx \\ dy \end{bmatrix} = \begin{bmatrix} \dfrac{\partial x}{\partial \xi} & \dfrac{\partial x}{\partial \eta} \\ \dfrac{\partial y}{\partial \xi} & \dfrac{\partial y}{\partial \eta} \end{bmatrix} \begin{bmatrix} d\xi \\ d\eta \end{bmatrix} \tag{5.2.2}$$

The 2×2 matrix of partial derivatives in (5.2.2) is called the *Jacobian matrix* of the transformation (5.2.1) and is denoted $\mathbf{J}$.

Equation (5.2.2) can be viewed as a linear transformation of line segments

$d\xi$ and $d\eta$ in $\hat{\Omega}$ into line segments dx and dy in the x, y-plane. If, at a point $(\xi, \eta) \in \hat{\Omega}$, it is possible to solve (5.2.2) for $d\xi$ and $d\eta$ in terms of dx and dy, then an *inverse map* T_e^{-1} of the x, y-coordinates into the ξ, η-coordinates can be constructed at this point. Obviously, a necessary and sufficient condition for the system (5.2.2) to be invertible is that the determinant $|\mathbf{J}|$ of the Jacobian matrix be nonzero at $(\xi, \eta) \in \hat{\Omega}$. The function $|\mathbf{J}|$ is called the *Jacobian* of the transformation (5.2.1),

$$|\mathbf{J}| = \det \mathbf{J} = \frac{\partial x}{\partial \xi}\frac{\partial y}{\partial \eta} - \frac{\partial x}{\partial \eta}\frac{\partial y}{\partial \xi} \tag{5.2.3}$$

Thus, whenever $|\mathbf{J}| \neq 0$, we can write

$$\begin{bmatrix} d\xi \\ d\eta \end{bmatrix} = \mathbf{J}^{-1}\begin{bmatrix} dx \\ dy \end{bmatrix} = \frac{1}{|\mathbf{J}|}\begin{bmatrix} \dfrac{\partial y}{\partial \eta} & -\dfrac{\partial x}{\partial \eta} \\ -\dfrac{\partial y}{\partial \xi} & \dfrac{\partial x}{\partial \xi} \end{bmatrix}\begin{bmatrix} dx \\ dy \end{bmatrix} \tag{5.2.4}$$

and

$$T_e^{-1}: \quad \begin{aligned} \xi &= \xi(x, y) \\ \eta &= \eta(x, y) \end{aligned} \Bigg\} \tag{5.2.5}$$

defines a map of element Ω_e back into the master element. Note that, as in (5.2.2),

$$\begin{bmatrix} d\xi \\ d\eta \end{bmatrix} = \begin{bmatrix} \dfrac{\partial \xi}{\partial x} & \dfrac{\partial \xi}{\partial y} \\ \dfrac{\partial \eta}{\partial x} & \dfrac{\partial \eta}{\partial y} \end{bmatrix}\begin{bmatrix} dx \\ dy \end{bmatrix} \tag{5.2.6}$$

Equating terms in (5.2.6) and (5.2.4), we get

$$\begin{aligned} \frac{\partial \xi}{\partial x} &= \frac{1}{|\mathbf{J}|}\frac{\partial y}{\partial \eta}, & \frac{\partial \xi}{\partial y} &= -\frac{1}{|\mathbf{J}|}\frac{\partial x}{\partial \eta}, \\ \frac{\partial \eta}{\partial x} &= -\frac{1}{|\mathbf{J}|}\frac{\partial y}{\partial \xi}, & \frac{\partial \eta}{\partial y} &= \frac{1}{|\mathbf{J}|}\frac{\partial x}{\partial \xi} \end{aligned} \tag{5.2.7}$$

These relationships will prove to be crucial in transforming calculations on $\hat{\Omega}$ into corresponding calculations for each element.

5.2.2 Construction of the Transformations T_e

We now come to the important issue of constructing the element transformations T_e. We are guided by the following criteria:

1.   Within each element, the functions $\xi = \xi(x, y)$ and $\eta = \eta(x, y)$ must

be invertible and continuously differentiable (else we could not use the convenient relations (5.2.7) in element calculations).

2. The sequence of mappings $\{T_e\}$ must generate a mesh with no spurious gaps between elements and with no element overlapping another.

3. Each map T_e should be easy to construct from the geometric data for the element.

4. The functions $x(\xi, \eta)$ and $y(\xi, \eta)$ defining the element maps should be easy to manipulate mathematically.

There are several ways in which one might construct maps T_e that satisfy the foregoing criteria. There is, however, a very natural means for determining a suitable map already at our disposal. This is based on the local interpolation ideas discussed in Section 4.4. Suppose that the shape functions ψ_j described in our discussion of interpolation are defined on a master element $\hat{\Omega}$. We recall that these shape functions can be used to construct an approximation $\hat{g}$ of any given function $g = g(\xi, \eta)$ by interpolating g at the nodal points,

$$\hat{g}(\xi, \eta) = \sum_{j=1}^{M} g_j \hat{\psi}_j(\xi, \eta) \tag{5.2.8}$$

where $g_j = g(\xi_j, \eta_j)$, (ξ_j, η_j) being the coordinates of node j, and where we have written $\hat{\psi}_j$ to emphasize that the shape functions now pertain to the master element. Obviously, M denotes the number of nodes defined in the master element.

With (5.2.8) in mind, recall the form of the transformation (5.2.1) and observe that the coordinate variables x and y may themselves be viewed simply as linear functions on Ω_e. Hence, *the finite element shape functions can be utilized to construct the map* T_e,

$$T_e: \left.\begin{array}{l} x = \sum_{j=1}^{M} x_j \hat{\psi}_j(\xi, \eta) \\[2mm] y = \sum_{j=1}^{M} y_j \hat{\psi}_j(\xi, \eta) \end{array}\right\} \tag{5.2.9}$$

Here, (x_j, y_j) are the x, y-coordinates of local nodal point j in element Ω_e.

Before continuing, let us check to see if the four criteria listed earlier are satisfied by transformations of the type (5.2.9). Criteria 3 and 4 are easily verified. The map T_e of (5.2.9) is readily constructed from element data (the coordinates $(x_1, y_1), (x_2, y_2), \ldots, (x_M, y_M)$), and since the shape functions are polynomials, the mapping functions are easy to manipulate and are continuously differentiable infinitely many times in $\hat{\Omega}$. Criterion 2 is usually not difficult to satisfy. Moreover, given the right connectivity of nodal points, the generation of continuous global basis functions ϕ_i is automatic in the process suggested by (5.2.9). For example, the quadratic shape functions on the master triangle and biquadratics on the master square shown in Fig. 5.3 map these elements to the corresponding elements Ω_e in the x, y-plane in such a way that straight sides of the master element are mapped to quadratic curved

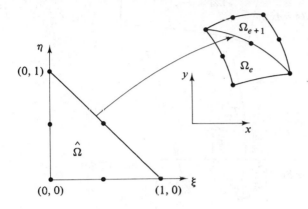

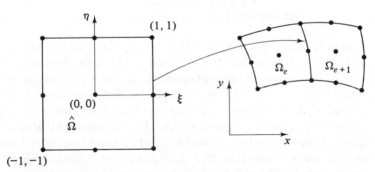

FIGURE 5.3 *Maps defined by quadratic shape functions and biquadratic shape functions on the master triangle and square, respectively. The quadratic curve between two elements in the x, y-plane is defined uniquely by the maps as shown.*

sides of Ω_e. On a given curved side between two adjacent elements Ω_e and Ω_{e+1}, the maps T_e and T_{e+1} reduce to the same quadratic functions since nodal-point coordinates defining the sides of both elements, and thus the maps for these sides, are the same. Hence, the interelement boundary is uniquely determined—no gaps between elements or overlapping can occur.

A geometric interpretation of the Jacobian is indicated in Fig. 5.4. At a

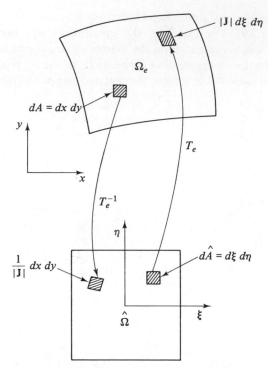

FIGURE 5.4 *Transformation of area elements.*

point (ξ, η) in $\hat{\Omega}$ an element of area $d\hat{A}$ is given by $d\hat{A} = d\xi\, d\eta$. The image of this area in the x, y-plane has the area $dA = |\mathbf{J}|\, d\xi d\eta$. The value of $|\mathbf{J}|$ can be viewed as the ratio of areas of elements at points $(x(\xi, \eta), y(\xi, \eta))$ and (ξ, η). Since both (x, y) and (ξ, η) are right-handed coordinate systems, a positive value of $|\mathbf{J}|$ at all points indicates that (5.2.9) is a *proper* transformation. The presence in $\hat{\Omega}$ of points at which $|\mathbf{J}| = 0$ indicates that area from $\hat{\Omega}$ is "squeezed" into a line (or a point) in Ω_e. This condition implies that the mapping T_e is *not invertible* (i.e., T_e^{-1} does not exist). Negative values of $|\mathbf{J}|$ indicate that some portion of $\hat{\Omega}$ has been "turned inside out" in being mapped into the x, y-plane.

Clearly, for the mapping defined by (5.2.5) to be acceptable for our purposes, we must have positive values of $|\mathbf{J}|$ at all points in $\hat{\Omega}$. The satisfaction of this condition is not assured in general for all maps of the form (5.2.9). Each set of shape functions must be examined and criteria developed which will guarantee that the choice of $\hat{\psi}_i$, x_i, and y_i ensures that $|\mathbf{J}| > 0$ throughout $\hat{\Omega}$.

Returning now to (5.2.7), suppose that all of our criteria for element transformations are satisfied by the map (5.2.9). The calculations in (5.2.7) now assume the form

$$\left.\begin{array}{ll}
\dfrac{\partial \xi}{\partial x} = \dfrac{1}{|\mathbf{J}|} \displaystyle\sum_{j=1}^{M} y_j \dfrac{\partial \hat{\psi}_j}{\partial \eta}, & \dfrac{\partial \xi}{\partial y} = -\dfrac{1}{|\mathbf{J}|} \displaystyle\sum_{j=1}^{M} x_j \dfrac{\partial \hat{\psi}_j}{\partial \eta} \\[4mm]
\dfrac{\partial \eta}{\partial x} = -\dfrac{1}{|\mathbf{J}|} \displaystyle\sum_{j=1}^{M} y_j \dfrac{\partial \hat{\psi}_j}{\partial \xi}, & \dfrac{\partial \eta}{\partial y} = \dfrac{1}{|\mathbf{J}|} \displaystyle\sum_{j=1}^{M} x_j \dfrac{\partial \hat{\psi}_j}{\partial \xi} \\[4mm]
|\mathbf{J}| = \left\{ \displaystyle\sum_{j=1}^{M} x_j \dfrac{\partial \hat{\psi}_j}{\partial \xi} \right\} \left\{ \displaystyle\sum_{j=1}^{M} y_j \dfrac{\partial \hat{\psi}_j}{\partial \eta} \right\} - \left\{ \displaystyle\sum_{j=1}^{M} x_j \dfrac{\partial \hat{\psi}_j}{\partial \eta} \right\} \left\{ \displaystyle\sum_{j=1}^{M} y_j \dfrac{\partial \hat{\psi}_j}{\partial \xi} \right\}
\end{array}\right\} \quad (5.2.10)$$

In Section 5.3 we show how these formulas can be used in constructing element matrices.

The following examples illustrate the ideas presented in this section.

Examples: Figure 5.5 shows a four-node master element $\hat{\Omega}$ and four elements Ω_e, $e = 1, 2, 3, 4$, generated from it using the map of (5.2.9). The shape functions $\hat{\psi}_i$ defined on the master element are (see Section 4.4)

$$\hat{\psi}_1(\xi, \eta) = \tfrac{1}{4}(1 - \xi)(1 - \eta) \qquad \hat{\psi}_2(\xi, \eta) = \tfrac{1}{4}(1 + \xi)(1 - \eta)$$
$$\hat{\psi}_3(\xi, \eta) = \tfrac{1}{4}(1 + \xi)(1 + \eta) \qquad \hat{\psi}_4(\xi, \eta) = \tfrac{1}{4}(1 - \xi)(1 + \eta)$$

In this example, straight lines $\xi = $ constant or $\eta = $ constant in $\hat{\Omega}$ map to corresponding straight lines in Ω_e. For example, the line $\eta = 0$ bisects the vertical sides of $\hat{\Omega}$ and is mapped to a straight line bisecting a pair of opposite sides of Ω_e. For this map to be invertible, it is necessary that no interior angle of Ω_e equal or exceed π. We shall examine the behavior of the Jacobian $|\mathbf{J}|$ for each of the elements shown.

Element Ω_1

In this case, the nodal-point coordinates are

$$(x_j, y_j) = \{(3, 0), (3, 1), (0, 1), (0, 0)\}$$

so the coordinate map is

$$T_1: \quad \begin{array}{l} x = 3\hat{\psi}_1 + 3\hat{\psi}_2 = \tfrac{3}{2}(1 - \eta) \\[2mm] y = \hat{\psi}_2 + \hat{\psi}_3 = \tfrac{1}{2}(1 + \xi) \end{array}$$

The Jacobian is

$$|\mathbf{J}| = \det \begin{bmatrix} 0 & -\tfrac{3}{2} \\[2mm] \tfrac{1}{2} & 0 \end{bmatrix} = \tfrac{3}{4}$$

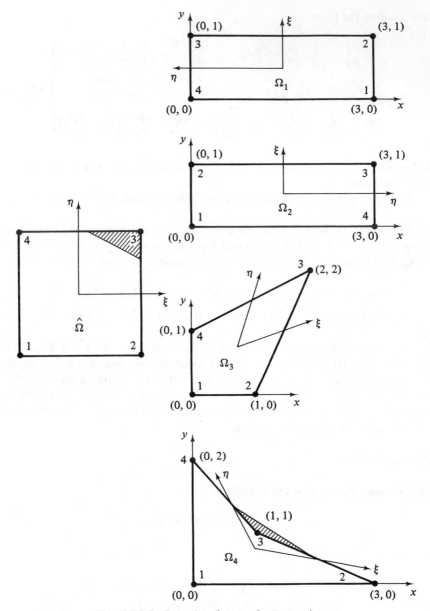

FIGURE 5.5 *Examples of maps of a master element.*

Since $|\mathbf{J}|$ is a positive constant, the map T_1 is invertible. In fact, all parts of $\hat{\Omega}$ are deformed in the same manner by a linear compression in the ξ direction and linear stretching in the η direction. The constant $|\mathbf{J}|$ is the ratio of the area of Ω_1 to the area of $\hat{\Omega}$.

Element Ω_2

Similar calculations for element Ω_2 reveal that

$$|\mathbf{J}| = -\tfrac{3}{4}$$

The clockwise ordering of nodal points has produced a negative Jacobian. This situation should be avoided.

Element Ω_3

In this case, the map is

$$T_3: \quad \begin{aligned} x &= \hat{\psi}_2 + 2\hat{\psi}_3 = \tfrac{1}{4}(3 + 3\xi + \eta + \zeta\eta) \\ y &= 2\hat{\psi}_3 + \hat{\psi}_4 = \tfrac{1}{4}(3 + \xi + 3\eta + \zeta\eta) \end{aligned}$$

so that

$$|\mathbf{J}| = \tfrac{1}{2} + \tfrac{1}{8}\xi + \tfrac{1}{8}\eta$$

Since the Jacobian is linear and is positive at each corner of $\hat{\Omega}$, it cannot be zero on $\hat{\Omega}$. Thus, the map is invertible. The value of $|\mathbf{J}|$ is smallest near node 1 and largest near node 3, indicating the relative stretching of different parts of $\hat{\Omega}$ by T_3.

Element Ω_4

For this element, we find that

$$|\mathbf{J}| = \tfrac{1}{8}(5 - 3\xi - 4\eta)$$

The Jacobian is *not* positive everywhere in $\hat{\Omega}$. In fact, $|\mathbf{J}| = 0$ along the line $\xi = \tfrac{5}{3} - \tfrac{4}{3}\eta$. At points above this critical line, $|\mathbf{J}| < 0$, and below it, $|\mathbf{J}| > 0$. The region above the line, shown in Fig. 5.5, is mapped *outside* of Ω_4 by T_4. Clearly, element Ω_4 is unacceptable. The trouble can be traced to the fact that the interior angle at node 3 is greater than π. It can be shown that for the four-node master element and the bilinear, tensor-product shape functions used in this example, T_e will be invertible if and only if all angles of the element are less than π.

The foregoing example indicates one type of restriction on the element data if $|\mathbf{J}|$ is not to vanish in the interior. Another restriction is encountered when side and interior nodes are introduced, as in the biquadratics of Fig. 5.3. The side nodes and also any interior node must lie within certain ranges if we are to preserve the basic invertibility condition for T_e. For example, each side node must be within the "middle half" of the curved side. (See Exercise 5.4.3.)

EXERCISES

5.2.1 Consider the region Ω of the x, y-plane shown below.

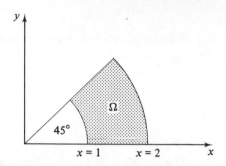

(a) Write equations defining each part of $\partial\Omega$ in terms of x and y.

(b) Calculate the integral of $x^2 + y^2$ over Ω using x and y as coordinate variables.

(c) Define and sketch a region $\hat{\Omega}$ of the r, θ-plane that is mapped onto Ω by the transformation

$$x = r \cos \theta$$
$$y = r \sin \theta$$

(d) Calculate the Jacobian of the coordinate transformation and show corresponding elements of area in Ω and $\hat{\Omega}$.

(e) Write down the inverse map and verify the relations (5.2.7).

(f) Repeat the calculation of part (b) using r and θ as coordinate variables.

5.2.2 For the four-node master element $\hat{\Omega}$ of Fig. 5.5 and the sets of nodal point coordinates given below complete the following:

(a) Sketch the elements Ω_e, showing the node numbers and the ξ- and η-axes.

(b) Sketch the image of the diagonal line $\xi = \eta$ in each element.

(c) Calculate the mapping functions and the Jacobians of the maps T_e. Check for invertibility of T_e.
Nodal coordinates:

(i) $(x_i, y_i) = \{(0, -1), (0, -2), (1, 2), (1, -1)\}$

(ii) $(x_i, y_i) = \{(0, 0), (1, -1), (1, 1), (0, 0)\}$

(iii) $(x_i, y_i) = \{(0, 0), (1, 0), (2, 0), (2, 1)\}$

5.2.3 Given the list of nodal points and their coordinates and the list of elements and their node numbers below:

(a) Sketch the finite element mesh Ω_h.

186

(b) Sketch the ξ, η-axes in each element.

(c) Verify that the maps T_e, $e = 1, 2, \ldots, 5$, produce a connected region Ω_h.

(d) Sketch the global basis function ϕ_4 for node 4 of the mesh.

Node	x	y	Element	Nodes
1	0	1	1	1, 2, 5, 4
2	0.7	0.7	2	3, 6, 5, 2
3	1	0	3	5, 8, 7, 4
4	0	2	4	5, 10, 9, 8
5	1.5	1.5	5	6, 11, 10, 5
6	2	0		
7	0	3		
8	1.5	3		
9	3	3		
10	3	1.5		
11	3	0		

5.2.4 For element Ω_3 of Fig. 5.5, calculate the values of $\partial \hat{\psi}_1 / \partial x$ and $\partial \hat{\psi}_2 / \partial y$ at the points $(0.5, 0.5)$ and $(0.5, 0)$ in $\hat{\Omega}$.

5.3 FINITE ELEMENT CALCULATIONS

Having established how acceptable sequences of coordinate maps T_e from $\hat{\Omega}$ into elements Ω_e can be constructed via (5.2.9), we turn to the fundamental question of how these ideas can be used in solving linear, second-order boundary-value problems of the type discussed in Chapter 4.

The key to the finite element approximation of such problems is the calculation of the element matrices for each element in the mesh. In particular, for the approximation of problem (4.2.18) (recall equations (4.5.10), (4.5.11), (4.5.20), and (4.5.21)), we must evaluate the integrals

$$k_{ij}^e = \int_{\Omega_e} \left[k \left(\frac{\partial \psi_i^e}{\partial x} \frac{\partial \psi_j^e}{\partial x} + \frac{\partial \psi_i^e}{\partial y} \frac{\partial \psi_j^e}{\partial y} \right) + b \psi_i^e \psi_j^e \right] dx \, dy \qquad (5.3.1a)$$

$$f_i^e = \int_{\Omega_e} f \psi_i^e \, dx \, dy \qquad (5.3.1b)$$

$$P_{ij}^e = \int_{\partial \Omega_{2h}^e} p \psi_i^e \psi_j^e \, ds \qquad (5.3.1c)$$

$$\gamma_i^e = \int_{\partial \Omega_{2h}^e} \gamma \psi_i^e \, ds \qquad (5.3.1d)$$

with $1 \leq e \leq E$ and $1 \leq i, j \leq N_e$. In (5.3.1a) and (5.3.1b), it is understood that the quantities appearing in the integrands are functions of the global Cartesian coordinates x and y; in (5.3.1c) and (5.3.1d), the integrands are functions of the parameter s defining the boundary $\partial\Omega_{2h}(x = x(s), y = y(s), (x, y) \in \partial\Omega_{2h})$.

Once the matrices (5.3.1) are available for each element, the remainder of the analysis proceeds as described in Section 4.6.2. In the remainder of this section, we shall describe how these matrices are computed using the master element as the setting for calculations and the element maps T_e of (5.2.9).

5.3.1 Master-Element Calculations

We begin by choosing a master element $\hat{\Omega}$ with coordinates ξ and η. It is at this stage that we decide upon the type of element to be used for our approximation. The geometry of $\hat{\Omega}$ is chosen to be as simple as possible to facilitate element calculations. For example, $\hat{\Omega}$ can be selected as the triangular or square elements introduced in Section 4.4. With a given choice of $\hat{\Omega}$, we have a corresponding set of master-element shape functions $\hat{\psi}_j$. For the moment, we shall assume that the same shape functions which define the element maps are used to define the approximations over each element; thus, if there are N_e nodes on each element, we take $M = N_e$ in (5.2.9). We shall select the set $\hat{\psi}_1, \hat{\psi}_2, \ldots, \hat{\psi}_{N_e}$, and the nodal point coordinates so that the maps T_e will always be invertible.

Having selected $\hat{\Omega}$ and $\hat{\psi}_j$, the remaining steps in our analysis proceed as follows.

1. **Element Maps:** Element coordinate maps T_e are constructed via (5.2.9),

$$T_e: \quad x = \sum_{j=1}^{N_e} x_j \hat{\psi}_j(\xi, \eta), \qquad y = \sum_{j=1}^{N_e} y_j \hat{\psi}_j(\xi, \eta) \qquad (5.3.2)$$

The element Ω_e to which T_e maps $\hat{\Omega}$ is completely determined by specifying the x, y-coordinates (x_j, y_j) of all the nodal points of Ω_e.

2. **Transformation of Shape Functions:** Let g be a scalar-valued function of x and y defined on an element Ω_e. Then we can convert g to a function $\hat{g}$ of ξ and η defined on $\hat{\Omega}$ by setting

$$g(x, y) = g(x(\xi, \eta), y(\xi, \eta)) = \hat{g}(\xi, \eta)$$

where $x(\xi, \eta)$ and $y(\xi, \eta)$ are now given by (5.3.2). Similarly, if $\hat{g}(\xi, \eta)$ is given on $\hat{\Omega}$, then under the transformation of (ξ, η) into (x, y),

$$\hat{g}(\xi, \eta) = \hat{g}(\xi(x, y), \eta(x, y)) = g(x, y) \qquad (5.3.3)$$

Thus, element shape functions $\psi_j^e = \psi_j^e(x, y)$ for Ω_e are simply obtained from $\hat{\psi}_j(\xi, \eta)$ by

$$\psi_j^e(x, y) = \hat{\psi}_j(\xi(x, y), \eta(x, y)), \qquad j = 1, 2, \ldots, N_e \qquad (5.3.4)$$

The derivatives of ψ_j^e are obtained by the chain rule:

$$\frac{\partial \psi_j^e}{\partial x} = \frac{\partial \hat{\psi}_j}{\partial \xi} \frac{\partial \xi}{\partial x} + \frac{\partial \hat{\psi}_j}{\partial \eta} \frac{\partial \eta}{\partial x}; \qquad \frac{\partial \psi_j^e}{\partial y} = \frac{\partial \hat{\psi}_j}{\partial \xi} \frac{\partial \xi}{\partial y} + \frac{\partial \hat{\psi}_j}{\partial \eta} \frac{\partial \eta}{\partial y} \qquad (5.3.5)$$

According to (5.3.2),

$$\left.\begin{array}{ll} \dfrac{\partial x}{\partial \xi} = \displaystyle\sum_{k=1}^{N_e} x_k \dfrac{\partial \hat{\psi}_k(\xi, \eta)}{\partial \xi}, & \dfrac{\partial x}{\partial \eta} = \displaystyle\sum_{k=1}^{N_e} x_k \dfrac{\partial \hat{\psi}_k(\xi, \eta)}{\partial \eta} \\[2ex] \dfrac{\partial y}{\partial \xi} = \displaystyle\sum_{k=1}^{N_e} y_k \dfrac{\partial \hat{\psi}_k(\xi, \eta)}{\partial \xi}, & \dfrac{\partial y}{\partial \eta} = \displaystyle\sum_{k=1}^{N_e} y_k \dfrac{\partial \hat{\psi}_k(\xi, \eta)}{\partial \eta} \end{array}\right\} \qquad (5.3.6)$$

Thus, using (5.2.7) and (5.3.6), equations (5.3.5) become

$$\left.\begin{array}{l} \dfrac{\partial \psi_j^e}{\partial x} = \dfrac{1}{|\mathbf{J}(\xi, \eta)|} \left\{ \dfrac{\partial \hat{\psi}_j}{\partial \xi} \displaystyle\sum_{k=1}^{N_e} y_k \dfrac{\partial \hat{\psi}_k}{\partial \eta} - \dfrac{\partial \hat{\psi}_j}{\partial \eta} \displaystyle\sum_{k=1}^{N_e} y_k \dfrac{\partial \hat{\psi}_k}{\partial \xi} \right\} \\[3ex] \dfrac{\partial \psi_j^e}{\partial y} = \dfrac{1}{|\mathbf{J}(\xi, \eta)|} \left\{ -\dfrac{\partial \hat{\psi}_j}{\partial \xi} \displaystyle\sum_{k=1}^{N_e} x_k \dfrac{\partial \hat{\psi}_k}{\partial \eta} + \dfrac{\partial \hat{\psi}_j}{\partial \eta} \displaystyle\sum_{k=1}^{N_e} x_k \dfrac{\partial \hat{\psi}_k}{\partial \xi} \right\} \end{array}\right\} \qquad (5.3.7)$$

where $|\mathbf{J}(\xi, \eta)|$ is the Jacobian of the transformation T_e. Note that the partial derivatives of ψ_j^e with respect to x and y are completely determined by calculations defined only on the master element $\hat{\Omega}$.

An important point arises in connection with the shape functions $\psi_j^e(x, y)$ computed via (5.3.4). Recall that in previous chapters we required that the functions $v_h^e(x, y) = \sum_{j=1}^{N_e} v_j \psi_j^e(x, y)$ contain complete polynomials of degree $k > 0$ and that, in fact, the rates of convergence of finite element interpolants depend upon k when the interpolated functions are sufficiently smooth. Now the transformed element shape functions may not be polynomials; for example, the coordinate transformations may lead to ratios of polynomials. Nevertheless, if the shape functions $\hat{\psi}_j$ for $\hat{\Omega}$ contain complete polynomials of degree k, and if the function $|\mathbf{J}|$ is well behaved (e.g., bounded above and below by positive constants), then the transformed functions ψ_j^e retain the same asymptotic convergence properties as complete polynomials. Thus, we can achieve the essential convergence properties by choosing appropriate shape functions for $\hat{\Omega}$ and by constructing smooth maps T_e with well-behaved Jacobians.

3. **Integration:** Let g and $\hat{g}$ be the functions defined in (5.3.3). Then it is clear that

$$\int_{\Omega_e} g(x, y) \, dx \, dy = \int_{\hat{\Omega}} \hat{g}(\xi, \eta) \, | \, \mathbf{J}(\xi, \eta)| \, d\xi \, d\eta \qquad (5.3.8)$$

Thus, integrals of functions over elements Ω_e can be evaluated using calculations on $\hat{\Omega}$ by this simple relationship. Notice that both $\hat{g}$ and $|\mathbf{J}|$ are determined by the element map (5.3.2); for example,

$$\hat{g}(\xi, \eta) = g\left(\sum_{j=1}^{N_\xi} x_j \hat{\psi}_j(\xi, \eta), \sum_{j=1}^{N_e} y_j \hat{\psi}_j(\xi, \eta) \right) \qquad (5.3.9)$$

In actual calculations, numerical quadrature formulas are generally used to evaluate all integrals. Recall that Gaussian quadrature rules were introduced in Chapter 3 to evaluate integrals for two-point boundary-value problems. Quadrature rules are typically defined for standard (geometrically simple) regions of integration, such as our master element, so that the form (5.3.8) is particularly convenient from this point of view.

In general, any quadrature formula (Gaussian or otherwise) is defined by specifying the coordinates (ξ_l, η_l) of a number N_l of *integration points* in the domain over which the integral is to be evaluated as well as a set of N_l numbers w_l called the *quadrature weights* $(l = 1, 2, \ldots, N_l)$. Thus, if $\hat{G}(\xi, \eta) = \hat{g}(\xi, \eta) \, |J(\xi, \eta)|$ is to be integrated over $\hat{\Omega}$, we use the formula

$$\int_{\Omega_e} g(x, y) \, dx \, dy = \int_{\hat{\Omega}} \hat{G}(\xi, \eta) \, d\xi \, d\eta = \sum_{l=1}^{N_l} \hat{G}(\xi_l, \eta_l) w_l + \hat{E} \qquad (5.3.10)$$

where $\hat{E}$ is the quadrature error.

If the integrand $\hat{G}(\xi, \eta)$ is a polynomial, the order N_l of the quadrature rule can always be chosen large enough that the integral is evaluated exactly. For example, if Gaussian quadrature is used for integrating a linear function over a master triangle, the integration will be exact if a one-point rule is employed with (ξ_1, η_1) the centroid and $w_1 = \frac{1}{2}$. Similarly, on the master square with biquadratic or bicubic integrands, the 2×2 Gauss rule with four quadrature points will produce an exact integration.

4. **Element Stiffnesses and Load Vectors:** The entries k_{ij}^e and f_i^e in the element stiffness matrix and load vector defined by (5.3.1a) and (5.3.1b) can now be computed directly using (5.3.7) and the integration rule (5.3.10). It is important to note that only the values of the integrands at the integration points in $\hat{\Omega}$ need be used in all of these

calculations. Specific equations for the calculation of k_{ij}^e and f_i^e are obtained by simple substitution (see Exercise 5.3.1).

If the modulus $k = k(x, y)$, the coefficient $b = b(x, y)$, and the source intensity $f = f(x, y)$ are not constant over an element, there is one additional calculation that is often made. Instead of k, b, and f in (5.3.1a) and (5.3.1b), we may use their interpolants,

$$k_h(x, y) = \sum_{j=1}^{N_e} k_j \psi_j^e(x, y), \qquad b_h(x, y) = \sum_{j=1}^{N_e} b_j \psi_j^e(x, y), \left.\vphantom{\sum_{j=1}^{N_e}}\right\}$$

$$f_h(x, y) = \sum_{j=1}^{N_e} f_j \psi_j^e(x, y) \tag{5.3.11}$$

where $k_j = k(x_j, y_j)$, etc., $j = 1, 2, \ldots, N_e$. Then the calculations of k_{ij}^e and f_i^e require only the nodal values of k, b, and f as data.

5. **Boundary Integrals:** The calculation of P_{ij}^e and γ_i^e of (5.3.1c) and (5.3.1d) is carried out by integrating along those sides of the master element that are mapped onto the sides $\partial\Omega_{2h}^e$ of Ω_e along which natural boundary conditions are prescribed. In any particular element, there may be no such sides or there may be one or more. We describe such calculations for a single side; the steps can be repeated when more than one side is to be treated. For definiteness, we suppose that the side $\xi = 1$ of a master square is to be mapped onto $\partial\Omega_{2h}^e$.

Let $\hat\theta_j$ denote the restriction of the master-element shape functions $\hat\psi_j$ to side $\xi = 1$:

$$\hat\theta_j(\eta) = \hat\psi_j(1, \eta), \qquad j = 1, 2, \ldots, N_e \tag{5.3.12}$$

It is clear that except for nodes j falling on the boundary $\xi = 1$, $\hat\theta_j = 0$. We then have

$$P_{ij}^e = \int_{\partial\Omega_{2h}^e} p\psi_i^e \psi_j^e \, ds = \int_{-1}^{1} \hat p(\eta)\hat\theta_i(\eta)\hat\theta_j(\eta)|j(\eta)| \, d\eta \left.\vphantom{\int_{-1}^1}\right\}$$

$$\gamma_i^e = \int_{\partial\Omega_{2h}^e} \gamma\psi_i^e \, ds = \int_{-1}^{1} \hat\gamma(\eta)\hat\theta_i(\eta)|j(\eta)| \, d\eta \tag{5.3.13}$$

where $|j|$ is the Jacobian of the transformation of η into the arc-length parameter s in the x, y-plane. Since

$$ds = \left\{ \left[\frac{\partial x\,(1, \eta)}{\partial\eta} \right]^2 + \left[\frac{\partial y\,(1, \eta)}{\partial\eta} \right]^2 \right\}^{1/2} d\eta$$

we see that

$$|j(\eta)| = \left\{ \left[\frac{\partial x\,(1, \eta)}{\partial\eta} \right]^2 + \left[\frac{\partial y\,(1, \eta)}{\partial\eta} \right]^2 \right\}^{1/2} \tag{5.3.14}$$

where $x(\xi, \eta)$ and $y(\xi, \eta)$ are defined in (5.3.2).

To evaluate the integrals in (5.3.13) over the side of the master element, we continue to use numerical integration formulas. In the present case, a one-dimensional quadrature rule of the type discussed in Chapter 3 is needed. Again, the functions p and γ in (5.3.13) can be replaced by their interpolants p_h and γ_h, if desired. With these conventions, the procedure is then essentially the same as that described for the evaluation of k_{ij}^e and f_i^e.

5.3.2 Computational Aspects

In anticipation of coding considerations to be taken up in Section 5.5, we now list the essential steps in the element matrix calculations. To evaluate k_{ij}^e and f_i^e, we carry out the following steps:

1. Choose $\hat{\Omega}$ and $\hat{\psi}_j$, $j = 1, 2, \ldots, N_e$, and specify the x, y-coordinates $(x_1, y_1), (x_2, y_2), \ldots, (x_{N_e}, y_{N_e})$ of nodal points of each element.

2. Specify a set of N_I integration points (ξ_l, η_l), $l = 1, 2, \ldots, N_I$, and quadrature weights w_l for $\hat{\Omega}$.

3. Calculate the values of $\hat{\psi}_j$, $\partial\hat{\psi}_j/\partial\xi$, $\partial\hat{\psi}_j/\partial\eta$ at the integration points (ξ_l, η_l) (see (5.3.4)).

4. Calculate the values of $x = x(\xi, \eta)$, $y = y(\xi, \eta)$ of (5.3.2) and their derivatives $\partial x/\partial\xi$, $\partial x/\partial\eta$, $\partial y/\partial\xi$, $\partial y/\partial\eta$, at the integration points (ξ_l, η_l) (see (5.3.6)).

5. Calculate the values of the Jacobian $|\mathbf{J}(\xi, \eta)|$ and the functions $\partial\xi/\partial x$, $\partial\xi/\partial y$, $\partial\eta/\partial x$, $\partial\eta/\partial y$ at the integration points (ξ_l, η_l) (see (5.2.7)).

6. Using (5.3.7) and the results of steps 3, 4, and 5, compute $\partial\psi_j^e/\partial x$ and $\partial\psi_j^e/\partial y$ at each integration point (ξ_l, η_l).

7. Calculate the values of k, b, and f (or k_h, b_h, and f_h) at the integration points (ξ_l, η_l).

8. Using the results of steps 3 through 7, calculate the values of the integrands in (5.3.1a), (5.3.1b) at the integration points (ξ_l, η_l) and multiply each by $w_l|\mathbf{J}(\xi_l, \eta_l)|$.

9. Sum the numbers computed in step 8 in accordance with (5.3.10) to obtain k_{ij}^e and f_i^e.

A similar sequence of calculations can be performed to yield the boundary integrals in (5.3.13). We leave these details as an exercise (Exercise 5.3.2).

As a final point, we note that shape functions used in defining the coordinate maps T_e of (5.3.2) need not be the same as those used in calculating the local approximate solution. For example, suppose that we identify M nodes and shape functions $\tilde{\psi}_j$ on $\hat{\Omega}$ to define the map

$$T_e: \quad x = \sum_{j=1}^{M} x_j \tilde{\psi}_j(\xi, \eta); \qquad y = \sum_{j=1}^{M} y_j \tilde{\psi}_j(\xi, \eta) \qquad (5.3.15)$$

whereas N_e nodes and shape functions $\hat{\psi}_k$, $k = 1, 2, \ldots, N_e$, are used to define the local shape functions for our approximation. Then polynomial maps for defining mesh geometry of higher, equal, or lower degree than those used in the actual approximation of u can be used. These choices of element maps are usually classified as follows:

$$\left.\begin{array}{ll} M > N_e: & T_e \text{ is a } superparametric \text{ map} \\ M = N_e: & T_e \text{ is an } isoparametric \text{ map} \\ M < N_e: & T_e \text{ is a } subparametric \text{ map} \end{array}\right\} \qquad (5.3.16)$$

Some examples of these choices are given in Fig. 5.6.

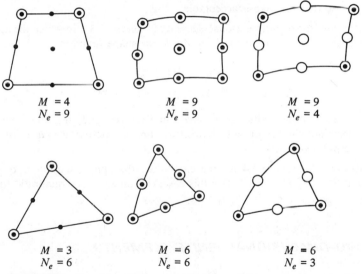

$$\begin{array}{ccc} M = 4 & M = 9 & M = 9 \\ N_e = 9 & N_e = 9 & N_e = 4 \end{array}$$

$$\begin{array}{ccc} M = 3 & M = 6 & M = 6 \\ N_e = 6 & N_e = 6 & N_e = 3 \end{array}$$

FIGURE 5.6 *Examples of subparametric, isoparametric, and superparametric elements.*

EXERCISES

5.3.1 Write out in detail specific formulas for the matrices in (5.3.1) in terms of element geometry and shape functions on the master element.

5.3.2 List the steps required to calculate the boundary integrals P_{ij}^e and γ_i^e in the spirit of the algorithm for k_{ij}^e and f_i^e itemized in Section 5.3.2.

5.3.3   Consider the one-dimensional element Ω_e (which may represent one side of a two-dimensional element) such that $0 \le x \le 1$. Let this element be the image of the master element $\hat{\Omega}$, for which $-1 \le \xi \le 1$, under the map constructed from the quadratic shape functions

$$\tilde{\theta}_1(\xi) = \tfrac{1}{2}\xi(\xi - 1)$$
$$\tilde{\theta}_2(\xi) = 1 - \xi^2$$
$$\tilde{\theta}_3(\xi) = \tfrac{1}{2}\xi(1 + \xi)$$

Note that the end nodes of Ω_e are located at $x_1 = 0$, $x_3 = 1$.

(a)   For each of the values of x_2 given below, plot the graph of the map T_e (i.e., plot x as a function of ξ).

$$x_2 = 0.2, \qquad x_2 = 0.3, \qquad x_2 = 0.5, \qquad x_2 = 0.8$$

(b)   Examine the Jacobian $dx/d\xi$ and establish bounds on the value of x_2 so that the map is acceptable.

5.3.4   Consider the one-dimensional element of Exercise 5.3.3 in which the approximation of u is constructed from the *linear* shape functions

$$\hat{\theta}_1(\xi) = \tfrac{1}{2}(1 - \xi)$$
$$\hat{\theta}_3(\xi) = \tfrac{1}{2}(1 + \xi)$$

For each of the values of x_2 given in Exercise 5.3.3, test this element to determine whether or not an arbitrary linear function $g = a + bx$ can be interpolated exactly.

5.3.5   Rework Exercise 5.3.4 for the case when the approximation is quadratic and $\hat{\theta}_i = \hat{\theta}_i$, $i = 1, 2, 3$. For this case, can an arbitrary quadratic function $g = a + bx + cx^2$ be interpolated exactly?

5.4   TWO-DIMENSIONAL FINITE ELEMENTS

The scheme for element calculations described in the previous section can be employed for many different kinds of finite elements. For a given variational problem such as the one defined by (4.3.9), the specification of an element type entails a prescription of the master element, the map T_e, the shape functions, and the quadrature rules. The information required to prescribe each of these items is given below.

1.   *Master element.* The geometry of the master element $\hat{\Omega}$ and the definitions of the element coordinate system ξ, η.

2.   *The map T_e.* The element coordinates (ξ_j, η_j) and shape functions $\hat{\psi}_j$

for the nodes, $j = 1, 2, \ldots , M$, that define the map from $\hat{\Omega}$ to Ω_e, through (5.3.15).

3. *Interpolation.* The element coordinates (ξ_j, η_j) and the shape functions $\hat{\psi}_j$ for the nodes $j = 1, 2, \ldots , N^e$, that determine the approximation $u_h^e = \sum_{j=1}^{N_e} u_j^e \psi_j^e$.

4. *Quadrature rules.* The number N_l of integration points, the element coordinates (ξ_l, η_l) and weights w_l for each integration point $l = 1, 2, \ldots , N_l$. Similar data are also required for the element boundary calculations.

In the following subsections, we give this information explicitly for quadrilateral elements (Section 5.4.1) and triangular elements (Section 5.4.2).

5.4.1 Quadrilateral Elements

Finite elements with four sides are generally referred to as quadrilateral elements, whether or not the sides are straight. The master element $\hat{\Omega}$ for quadrilateral elements is the familiar square used in the examples of Section 5.2. The element coordinate system ξ, η is Cartesian with the origin $(0, 0)$ located at the center of $\hat{\Omega}$.

The location of nodal points and the polynomial terms contained in the shape functions for the *Lagrange* family of quadrilateral elements were discussed in Chapter 4. The actual construction of shape functions for the Lagrange elements is described here. The key is to find functions containing polynomials of the proper degree that vanish at the right points. Shape functions that correspond to continuous global basis functions must satisfy the following condition: *the shape function $\hat{\psi}_i$ for node i must assume a unit value at node i and vanish at all other nodes of the element and along all sides of the element that do not contain node i.*

In order to construct a shape function containing polynomial terms of degree k, one simply takes the product of k linear forms, at least one of which vanishes at each point where the shape function must be zero. This product is then normalized so that the value at the associated node is 1. This procedure is clarified by sketches of the master element such as the ones shown in Fig. 5.7. For the four-node element, the bilinear shape functions contain (incomplete) polynomials in ξ and η of degree 2 and less. Each shape function should be written as the product of two linear forms. The shape function $\hat{\psi}_1$, for example, must vanish along sides $\xi = 1$ and $\eta = 1$, so we take the product $(1 - \xi)(1 - \eta)$. The value of this product at node 1 ($\xi = -1, \eta = -1$) is 4. The shape function $\hat{\psi}_1$ is thus $\hat{\psi}_1(\xi, \eta) = \frac{1}{4}(1 - \xi)(1 - \eta)$.

In the nine-node biquadratic element of Fig. 5.7, the shape functions contain some polynomials of degree 4. Shape function $\hat{\psi}_1$, which must vanish

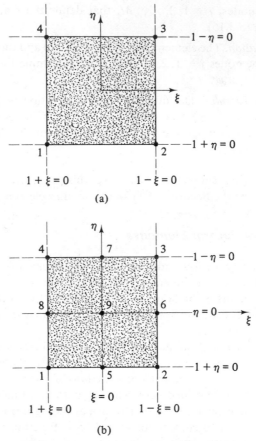

FIGURE 5.7 *Node location and equations of lines used to construct shape functions for (a) bi-linear and (b) bi-quadratic elements.*

along sides 2–3 and 3–4 and at nodes 5, 8, and 9, is composed of the product of the equations of the four lines which together contain all of these points. This product is $(\xi)(1 - \xi)(\eta)(1 - \eta)$, so that $\hat{\psi}_1(\xi, \eta) = \frac{1}{4}(\xi - \xi^2)(\eta - \eta^2)$. Similarly, we construct, $\hat{\psi}_6(\xi, \eta) = \frac{1}{2}(\xi)(1 + \xi)(1 - \eta)(1 + \eta) = \frac{1}{2}(\xi + \xi^2)(1 - \eta^2)$. This same procedure can be used to construct shape functions for all of the tensor-product elements.

There are useful shape functions for quadrilateral elements which are not constructed using tensor products of polynomials. A widely used class of quadrilateral elements is based on nodal configurations and shape functions in which no interior nodes (such as node 9 in Fig. 5.4b) appear.* The nodal

* The shape functions have been called "serendipity" functions, implying, perhaps facetiously, inadvertant good fortune associated with their discovery.

configuration for the eight-node quadrilateral element is shown in Fig. 5.8. The shape function for node 1 is constructed so as to vanish along the lines $1 - \xi = 0$, $1 - \eta = 0$, and $1 + \xi + \eta = 0$. Forming the product and nor-

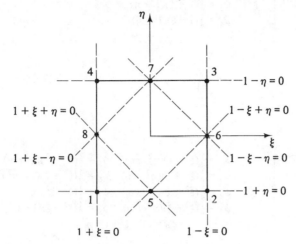

FIGURE 5.8 *Nodal location and equation of lines used to construct shape functions for quadratic eight-node element.*

malizing gives $\hat{\psi}_1(\xi, \eta) = \frac{1}{4}(1 - \xi)(1 - \eta)(-1 - \xi - \eta)$. The complete sets of shape functions for the four-, eight-, and nine-node elements are given in Fig. 5.9.

The shape functions for both the eight- and nine-node elements contain complete polynomials of second-degree plus the third-degree terms $x^2 y$ and $y^2 x$. The nine-node shape functions contain, in addition, the fourth-degree term $x^2 y^2$. As noted in Chapter 4, these terms of degree higher than 2 do not add to the rate of convergence because this is determined by the highest degree of complete polynomials in the $\hat{\psi}_j$ (i.e., second degree). The difference between the accuracy obtained using eight- and nine-node elements is, in fact, small and the choice between these elements is usually based on considerations such as ease of coding and resulting bandwidth.

The extension of the tensor product and serendipity families to elements containing complete cubic (or higher) polynomials is straightforward (see Exercise 5.4.2). Elements of higher degree than quadratic are seldom used in applications.

For all of the quadrilateral elements of the types discussed here, the restriction of the shape functions to element boundaries produces complete polynomials (in one dimension) of the same degree as the complete polynomial reproducible in the element. Thus, the four-node element has complete linear polynomials along its boundary, and the eight- and nine-node elements have complete quadratics. Hence, the shape functions $\hat{\theta}_j$ used in

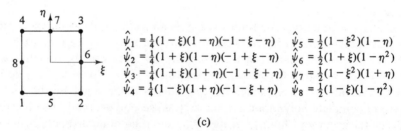

$$\hat{\psi}_1 = \tfrac{1}{4}(1 - \xi)(1 - \eta)$$
$$\hat{\psi}_2 = \tfrac{1}{4}(1 + \xi)(1 - \eta)$$
$$\hat{\psi}_3 = \tfrac{1}{4}(1 + \xi)(1 + \eta)$$
$$\hat{\psi}_4 = \tfrac{1}{4}(1 - \xi)(1 + \eta)$$

(a)

$$\hat{\psi}_1 = \tfrac{1}{4}(\xi^2 - \xi)(\eta^2 - \eta) \qquad \hat{\psi}_5 = \tfrac{1}{2}(1 - \xi^2)(\eta^2 - \eta)$$
$$\hat{\psi}_2 = \tfrac{1}{4}(\xi^2 + \xi)(\eta^2 - \eta) \qquad \hat{\psi}_6 = \tfrac{1}{2}(\xi^2 + \xi)(1 - \eta^2)$$
$$\hat{\psi}_3 = \tfrac{1}{4}(\xi^2 + \xi)(\eta^2 + \eta) \qquad \hat{\psi}_7 = \tfrac{1}{2}(1 - \xi^2)(\eta^2 + \eta)$$
$$\hat{\psi}_4 = \tfrac{1}{4}(\xi^2 - \xi)(\eta^2 + \eta) \qquad \hat{\psi}_8 = \tfrac{1}{2}(\xi^2 - \xi)(1 - \eta^2)$$
$$\hat{\psi}_9 = (1 - \xi^2)(1 - \eta^2)$$

(b)

$$\hat{\psi}_1 = \tfrac{1}{4}(1 - \xi)(1 - \eta)(-1 - \xi - \eta) \qquad \hat{\psi}_5 = \tfrac{1}{2}(1 - \xi^2)(1 - \eta)$$
$$\hat{\psi}_2 = \tfrac{1}{4}(1 + \xi)(1 - \eta)(-1 + \xi - \eta) \qquad \hat{\psi}_6 = \tfrac{1}{2}(1 + \xi)(1 - \eta^2)$$
$$\hat{\psi}_3 = \tfrac{1}{4}(1 + \xi)(1 + \eta)(-1 + \xi + \eta) \qquad \hat{\psi}_7 = \tfrac{1}{2}(1 - \xi^2)(1 + \eta)$$
$$\hat{\psi}_4 = \tfrac{1}{4}(1 - \xi)(1 + \eta)(-1 - \xi + \eta) \qquad \hat{\psi}_8 = \tfrac{1}{2}(1 - \xi)(1 - \eta^2)$$

(c)

FIGURE 5.9 *Master element, node numbering and shape functions for 4-, 9-, and 8-node quadrilateral finite elements.*

boundary integral calculations are exactly the one-dimensional shape functions defined in Chapter 2.

Quadrature rules for quadrilateral elements are usually derived from one-dimensional quadrature formulas by treating the integration over the master element as a double integral. If we write

$$\int_{\hat{\Omega}} G(\xi, \eta) \, d\xi \, d\eta = \int_{-1}^{1} \left[\int_{-1}^{1} G(\xi, \eta) \, d\xi \right] d\eta$$

and approximate both the integration with respect to ξ and with respect to η

by one-dimensional quadrature rules of order N such as those discussed in Chapter 3, we get

$$\int_{\hat{\Omega}} G(\xi, \eta) \, d\xi \, d\eta \simeq \sum_{m=1}^{N} \left[\sum_{n=1}^{N} G(\xi_n, \eta_m) w_n \right] w_m \qquad (5.4.1)$$

The double summation in (5.4.1) can be reduced to a single summation over the entire set of integration points by relabeling the point (m, n) with the single index l with $N_l = N^2$,

$$\sum_{m=1}^{N} \left[\sum_{n=1}^{N} G(\xi_n, \eta_m) w_n \right] w_m = \sum_{l=1}^{N^2} G(\xi_l, \eta_l) \bar{w}_l \qquad (5.4.2)$$

The corresponding labels of the integration points for Gauss integration of order 3 are shown in Fig. 5.10. Note that when the integral is written as a single sum the weights $\bar{w}_l$ are products of weights from the one-dimensional rule (see subroutine **SETINT** in **CODE1** of Chapter 3).

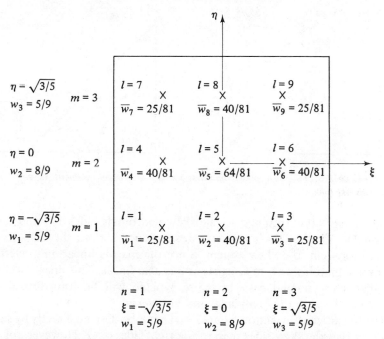

FIGURE 5.10 *Nine-point quadrature rule integration points and weights as a tensor product of three point quadratures with respect to ξ and η.*

5.4.2 Triangular Elements

The triangle is the simplest two-dimensional polygon in the sense that it has the fewest sides and vertices. As noted in Chapter 4, it is easy to choose exactly enough nodes in a triangular element to define shape functions which are complete polynomials of any specified degree (see Fig. 4.11). In our present study of triangular finite elements, we begin by considering the case of triangles Ω_e that have straight sides. The map from the right-isosceles master triangle is then linear, and both the analysis of the map and the formulation of the element matrices simplify accordingly. The natural generalization to triangles with curved sides is developed in a later part of this subsection.

As in the treatment of the quadrilateral element, the master element and map are fundamental to the development of the method for triangles. A simple linear transformation maps the right-isosceles master triangle $\hat{\Omega}$ onto an element Ω_e, as indicated in Fig. 5.11. The master element coordinates lines

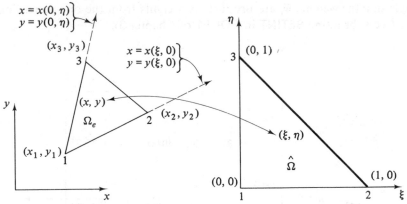

FIGURE 5.11   *Linear map from triangle Ω_e to master element $\hat{\Omega}$ and inverse map.*

$\xi = 0$ and $\eta = 0$ correspond to the skew coordinate sides 1–3 and 1–2 of element Ω_e. That is, the linear transformation describes the map from a Cartesian system to a skew system. Since the map is linear and invertible, the inverse map from Ω_e to $\hat{\Omega}$ exists and is also linear. This linearity implies also that a polynomial basis in the x, y-plane will be transformed to a polynomial basis in the ξ, η-plane, and vice versa.

The linear transformation from Ω_e to $\hat{\Omega}$ can be derived directly by simply requiring that the skew sides map to the right sides of $\hat{\Omega}$. However, for consistency, we shall again use the parametric map from the master element. The three linear shape functions for the right-isosceles master element $\hat{\Omega}$ can be written down by inspection, since each must assume a unit value at a corresponding vertex and be zero on the side opposite that vertex. They are

$$\hat{\psi}_1(\xi, \eta) = 1 - \xi - \eta, \qquad \hat{\psi}_2(\xi, \eta) = \xi, \qquad \hat{\psi}_3(\xi, \eta) = \eta \quad (5.4.3)$$

for nodes 1, 2, and 3, respectively. The coordinate map T_e is then defined by

$$x = \sum_{j=1}^{3} x_j \hat{\psi}_j(\xi, \eta), \qquad y = \sum_{j=1}^{3} y_j \hat{\psi}_j(\xi, \eta) \qquad (5.4.4)$$

where (x_j, y_j), $j = 1, 2, 3$, are the x, y-coordinates of the vertices in the local numbering system.

Introducing (5.4.3) into (5.4.4) and inverting, we obtain for the map T_e^{-1} the relations

$$\xi = \frac{1}{2A_e}[(y_3 - y_1)(x - x_1) - (x_3 - x_1)(y - y_1)]$$

$$\eta = \frac{1}{2A_e}[-(y_2 - y_1)(x - x_1) + (x_2 - x_1)(y - y_1)] \qquad (5.4.5)$$

where A_e is the area of Ω_e. The right-hand sides of equations (5.4.5) are recognized as the linear shape functions of equation (4.4.7):

$$\xi = \psi_2^e(x, y), \qquad \eta = \psi_3^e(x, y), \qquad 1 - \xi - \eta = \psi_1^e(x, y) \quad (5.4.6)$$

It is straightforward to use these results to determine the derivatives $\partial\xi/\partial x$, $\partial\xi/\partial y$, $\partial\eta/\partial x$, $\partial\eta/\partial y$ of the coordinate functions ξ and η, the Jacobian $|\mathbf{J}|$, and then the related stiffness and load matrices.

Area Coordinates: The expressions ξ, η, and $1 - \xi - \eta$ in (5.4.5) and (5.4.6) may be conveniently interpreted in terms of area ratios. This leads to the introduction of *area coordinates* which facilitate both the development of families of shape functions for higher-degree approximation and also the derivation of formal integration formulas.

Let (ξ, η) be the coordinates of an arbitrary point in $\hat{\Omega}$ and (x, y) the corresponding point in Ω_e. Suppose that we construct straight lines joining (ξ, η) and (x, y) to the vertices, and denote by $\hat{a}_i$ and a_i the subtriangle areas opposite node i in $\hat{\Omega}$ and Ω_e, respectively. The terms $1 - \xi - \eta$, ξ, and η in (5.4.6) are easily seen to be the area ratios $\hat{a}_i/\hat{A}$, where $\hat{A} = \frac{1}{2}$ is the area of the master element. We define the area coordinates on $\hat{\Omega}$ as

$$\zeta_i = \frac{\hat{a}_i}{\hat{A}}, \qquad i = 1, 2, 3 \qquad (5.4.7)$$

or

$$\zeta_1 = 1 - \xi - \eta, \qquad \zeta_2 = \xi, \qquad \zeta_3 = \eta \qquad (5.4.8)$$

Since the transformation T_e is linear, the Jacobian $|\mathbf{J}|$ is constant and is the ratio of the areas of Ω_e and $\hat{\Omega}$. This implies that the map distorts areas uniformly. Thus, the meaning of area coordinates is preserved under linear transformations and we have

$$\zeta_i = \frac{\hat{a}_i}{\hat{A}} = \frac{a_i}{A_e}, \qquad i = 1, 2, 3 \tag{5.4.9}$$

The subtriangles and coordinate lines are indicated in Fig. 5.12. It is important to note that this property is *not* preserved if the map T_e is nonlinear: in subsequent discussion of triangles Ω_e with curved sides, our interpretation of area coordinates will apply only to area ratios in $\hat{\Omega}$.

On examining the subtriangle areas and coordinate lines in the figure, we observe the following:

1. Let the ζ_i-coordinates of an arbitrary interior point P be $(\alpha_1, \alpha_2, \alpha_3)$. The coordinate line $\zeta_i = \alpha_i = $ constant is parallel to the side of the

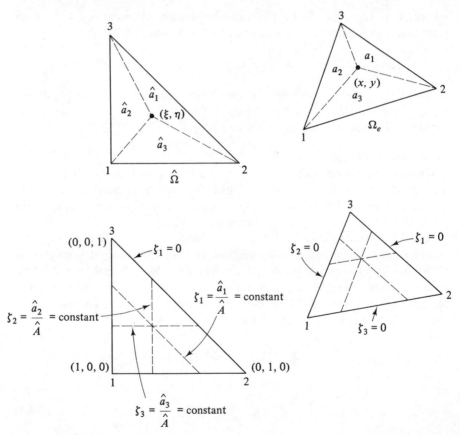

FIGURE 5.12 *Subtriangle areas* $\hat{a}_i$, a_i, *i = 1, 2, 3, and area coordinates* $\zeta_i = \hat{a}_i/\hat{A}$.

triangle opposite to vertex i: $\zeta_1 = \alpha_1$ is a line parallel to side 2–3, $\zeta_2 = \alpha_2$ is parallel to 1–2, and $\zeta_3 = \alpha_3$ is parallel to 1–2.

2. The coordinate lines $\zeta_i = 0$ define the boundary of the triangle: $\zeta_1 = 0$ is the side 2–3 opposite vertex 1, $\zeta_2 = 0$ is the side 1–3 opposite vertex 2, and $\zeta_3 = 0$ is the side 1–2 opposite vertex 3.

3. The triangular coordinates of the vertices 1, 2, and 3 are $(1, 0, 0)$, $(0, 1, 0)$, and $(0, 0, 1)$, respectively.

These properties will prove to be very useful in constructing shape functions of higher degree on triangles.

Higher-Degree Shape Functions: We now generalize the discussion to include nonlinear transformations T_e from $\hat{\Omega}$ to Ω_e. As stated above, the interpretation of coordinates ζ_i as area ratios is now not as useful on Ω_e since $|\mathbf{J}|$ is no longer constant. However, the triangular coordinates ζ_i on $\hat{\Omega}$ may be used to determine higher-degree shape functions $\hat{\psi}_i$ with ease. Concerning this point, notice that the linear shape functions on $\hat{\Omega}$ (Fig. 5.13a) are simply

$$\hat{\psi}_i(\zeta_1, \zeta_2, \zeta_3) = \zeta_i, \qquad i = 1, 2, 3 \tag{5.4.10}$$

For the node locations shown in Fig. 5.13b, the quadratic shape function for node 1 must be zero on the side $\zeta_1 = 0$ opposite node 1 and on $\zeta_1 = \frac{1}{2}$, which passes through the midside nodes 4 and 6. Scaling the product of these linear forms by 2 to ensure that $\hat{\psi}_1(1, 0, 0) = 1$ gives the quadratic shape function $\hat{\psi}_1(\zeta_1, \zeta_2, \zeta_3) = \zeta_1(2\zeta_1 - 1)$. Similar constructions yield the remaining shape functions, and we have

$$\hat{\psi}_i(\zeta_1, \zeta_2, \zeta_3) = \begin{cases} \zeta_i(2\zeta_i - 1), & \text{vertex node } i = 1, 2, 3 \\ 4\zeta_j\zeta_k, & \text{node } i \text{ midside } j\text{–}k \text{ for} \\ & i = 4, 5, 6 \end{cases} \tag{5.4.11}$$

In the same way, for a cubic Lagrange element with nodes located as in Fig. 5.13c, the shape functions are

$$\hat{\psi}_i(\zeta_1, \zeta_2, \zeta_3) = \begin{cases} \frac{9}{2}\zeta_i(\zeta_i - \frac{2}{3})(\zeta_i - \frac{1}{3}), & \text{vertex node } i, \, i = 1, 2, 3 \\ \frac{27}{2}\zeta_j\zeta_k(\zeta_j - \frac{1}{3}), & \text{first trisection} \\ & \text{node } (i = 4, 5, 6) \text{ side } j\text{–}k \\ \frac{27}{2}\zeta_j\zeta_k(\zeta_k - \frac{1}{3}), & \text{second trisection} \\ & \text{node } (i = 7, 8, 9) \text{ side } j\text{–}k \\ 27\zeta_1\zeta_2\zeta_3, & \text{centroid node } i = 10 \end{cases} \tag{5.4.12}$$

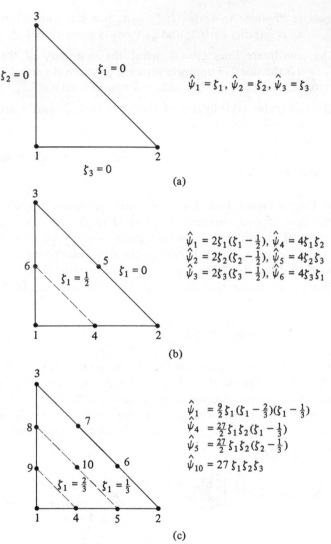

$$\hat{\psi}_1 = \zeta_1, \quad \hat{\psi}_2 = \zeta_2, \quad \hat{\psi}_3 = \zeta_3$$

(a)

$$\hat{\psi}_1 = 2\zeta_1(\zeta_1 - \tfrac{1}{2}), \quad \hat{\psi}_4 = 4\zeta_1\zeta_2$$
$$\hat{\psi}_2 = 2\zeta_2(\zeta_2 - \tfrac{1}{2}), \quad \hat{\psi}_5 = 4\zeta_2\zeta_3$$
$$\hat{\psi}_3 = 2\zeta_3(\zeta_3 - \tfrac{1}{2}), \quad \hat{\psi}_6 = 4\zeta_3\zeta_1$$

(b)

$$\hat{\psi}_1 = \tfrac{9}{2}\zeta_1(\zeta_1 - \tfrac{2}{3})(\zeta_1 - \tfrac{1}{3})$$
$$\hat{\psi}_4 = \tfrac{27}{2}\zeta_1\zeta_2(\zeta_1 - \tfrac{1}{3})$$
$$\hat{\psi}_5 = \tfrac{27}{2}\zeta_1\zeta_2(\zeta_2 - \tfrac{1}{3})$$
$$\hat{\psi}_{10} = 27\,\zeta_1\zeta_2\zeta_3$$

(c)

FIGURE 5.13 *Generation of shape functions $\hat{\psi}_i(\zeta)$ on a triangle for (a) linears, (b) quadratics, and (c) cubics.*

These results can be extended readily to higher-degree Lagrange triangles by using the nodal distributions for the Pascal triangle in Fig. 4.11a.

Coordinate maps T_e may be constructed using the shape functions $\hat{\psi}_i$ generated in this manner: $x(\xi, \eta) \equiv x(\zeta_1, \zeta_2, \zeta_3) = \sum_{j=1}^{M} x_j\hat{\psi}_j(\zeta_1, \zeta_2, \zeta_3)$, where, for example, $M = 6$ and $M = 10$ for quadratic and cubic maps defined by $\hat{\psi}_i$ in (5.4.11) and (5.4.12). In a similar manner, these representations of the master element shape functions can be utilized to approximate the solution and interpolate data on $\hat{\Omega}$. The details differ from those for the general master element described in Sections 5.2 and 5.3 only through the presence of the (redundant) area coordinate $\zeta_1 = 1 - \zeta_2 - \zeta_3$. For example, the chain rule for calculating derivatives of $\hat{\psi}_i$ becomes

$$\frac{\partial \hat{\psi}_i}{\partial x} = \frac{\partial \hat{\psi}_i}{\partial \zeta_1} \frac{\partial \zeta_1}{\partial x} + \frac{\partial \hat{\psi}_i}{\partial \zeta_2} \frac{\partial \zeta_2}{\partial x} + \frac{\partial \hat{\psi}_i}{\partial \zeta_3} \frac{\partial \zeta_3}{\partial x} \qquad (5.4.13)$$

Of course, having determined $\hat{\psi}_i$ in the manner described above, one can also write $\zeta_1 = 1 - \zeta_2$ ζ_3 with $\zeta_2 \equiv \xi$ and $\zeta_3 \equiv \eta$ to obtain $\hat{\psi}_i(\xi, \eta)$. The remaining calculations proceed precisely as in the general treatment of master element computations in Section 5.3.

Integration: When the element Ω_e has straight sides, the integrands in the stiffness and load calculations on $\hat{\Omega}$ are often polynomials. Convenient formal integration formulas can then be utilized (Exercise 5.4.8). In more general cases, the element integrals can be evaluated approximately using numerical quadrature. For polynomial integrands, these quadrature formulas can be chosen to evaluate the integrals exactly.

The boundary integral contributions again reduce to one-dimensional quadratures and can be treated using the Gaussian quadrature rules discussed in Chapter 3. The integrals on $\hat{\Omega}$ have the form, $\int_{\hat{\Omega}} G(\xi, \eta) \, d\xi \, d\eta$ or, equivalently, in terms of area coordinates

$$\int_{\hat{\Omega}} G(\zeta_1, \zeta_2, \zeta_3) \, d\zeta_2 \, d\zeta_3 \qquad (\zeta_1 = 1 - \zeta_2 - \zeta_3)$$

We shall consider only the latter representation and write the quadrature formula in standard form as

$$\int_{\hat{\Omega}} G(\zeta_1, \zeta_2, \zeta_3) \, d\zeta_2 \, d\zeta_3 \simeq \sum_{l=1}^{N_l} w_l G(\zeta_l) \qquad (5.4.14)$$

where $\zeta_l = ((\zeta_1)_l, (\zeta_2)_l, (\zeta_3)_l)$ are the area coordinates of the quadrature points in $\hat{\Omega}$, w_l are the quadrature weights, and N_l is the order of the quadrature formula. Sample quadrature formulas are summarized in Fig. 5.14. The "degree" of the rule indicated in the table implies that polynomials up to this degree are integrated exactly by the formula.

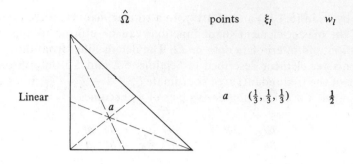

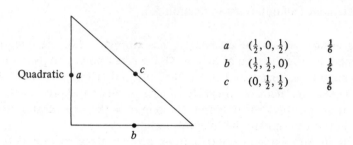

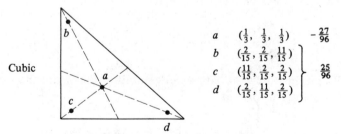

FIGURE 5.14   *Some quadrature rules for triangular elements.*

EXERCISES

5.4.1   Recall that the one-dimensional Gaussian quadrature formula with n points will integrate exactly polynomials up to degree $2n - 1$. For the case of constant coefficients and rectangular elements, the integrands in the stiffness calculation for quadrilateral elements are polynomials in x and y. For such a case, determine the number of integration points required to integrate the following terms exactly for the four-, eight-, and nine-node quadrilaterals:

$$\int_{\Omega_e} \left(\frac{\partial \psi_i^e}{\partial x} \frac{\partial \psi_j^e}{\partial x} + \frac{\partial \psi_i^e}{\partial y} \frac{\partial \psi_j^e}{\partial y} \right) dx\, dy$$

and

$$\int_{\Omega_e} \psi_i^e \psi_j^e \, dx\, dy$$

5.4.2 Verify that the nine-node element shape functions can represent exactly any function that the eight-node functions can and that the eight-node functions contain the four-node functions as special cases. Find a function that is represented exactly by the nine-node shape functions but not by the eight-node functions.

5.4.3 Show that the maps T_e obtained by using the four-, eight-, and nine-node shape functions will be identical when the coordinates of nodes 5 through 9 are chosen appropriately.

[*HINT:* Choose straight sides with midside nodes.]

5.4.4 Calculate the stiffness matrix for a square four-node element of dimension h. Use constant coefficients $k = k_0$ and $b = b_0$.

5.4.5 Using (5.3.1a) with $b(x, y) \equiv 0$, show that the stiffness matrix for an arbitrary triangular element over which $k(x, y) = k_0 = $ constant and linear shape functions are used, is given by

$$\mathbf{k}^e = \frac{k_0}{4A_e}(\mathbf{Y} + \mathbf{X})$$

where

$$\mathbf{Y} = \begin{bmatrix} (y_2 - y_3)^2 & (y_2 - y_3)(y_3 - y_1) & (y_2 - y_3)(y_1 - y_2) \\ (y_2 - y_3)(y_3 - y_1) & (y_3 - y_1)^2 & (y_3 - y_1)(y_1 - y_2) \\ (y_2 - y_3)(y_1 - y_2) & (y_3 - y_1)(y_1 - y_2) & (y_1 - y_2)^2 \end{bmatrix}$$

and $\mathbf{X}$ is the matrix obtained by replacing y_i by x_i in the definition of $\mathbf{Y}$.

5.4.6 Use the quadratic shape functions $\tilde{\psi}_i$ in equation (5.4.11) to construct the map T_e from $\hat{\Omega}$ to an element Ω_e that has quadratically curved sides. In many instances only one curved side is needed to model the boundary shape: develop the map from $\hat{\Omega}$ to such an element.

5.4.7 Verify that, in the example of Exercise 5.4.6, the interior node on the curved side has to lie in the "middle half" of the side (Compare with Exercise 5.3.3).

5.4.8 By direct evaluation on $\hat{\Omega}$ and side 1–2 of $\hat{\Omega}$, respectively, determine the formal integration formulas

$$\int_{\hat{\Omega}} \zeta_1^\mu \zeta_2^\nu \zeta_3^\tau \, d\zeta_2\, d\zeta_3 = \frac{\mu!\, \nu!\, \tau!}{(\mu + \nu + \tau + 2)!}$$

and

$$\int_0^1 \zeta_1^\mu \zeta_2^\nu \, d\zeta_2 = \frac{\mu! \, \nu!}{(\mu + \nu + 1)!}$$

Use these results to develop integration formulas for the same integrands on Ω_e and side 1–2 of Ω_e.

5.5 CODING OF TWO-DIMENSIONAL FINITE ELEMENT CALCULATIONS

The structure of the finite element program CODE1, described in Chapter 3, is sufficiently general to be used as a guide for two (or even three)-dimensional codes. For many practical applications, the large number of elements and nodal points required in problems of dimension higher than one preclude the use of the simple, in-core, data storage scheme and equation-solving algorithm used in CODE1. It is, however, possible to use these schemes in constructing a simple two-dimensional program that contains the essential logical components for the finite element solution of the class of boundary-value problems defined by equation (4.2.18).

In this section, we discuss modifications of CODE1 that will suffice to produce a simple two-dimensional program, CODE2. Except for some details to be discussed below, the flowchart in Fig. 3.2 describes the structure of CODE2 as well as CODE1. We give some coding details which, together with the structured set of programming assignments at the end of this section, will allow the reader to develop a working prototype of a two-dimensional finite element program.

5.5.1 Description of CODE2

The two-dimensional finite element program CODE2 treats boundary-value problems of the type defined by (4.2.18) in which the coefficients k and b, the load f, and the boundary data $\hat{u}$, $\hat{\sigma}$, and p are piecewise constants. The element library consists of quadratic isoparametric quadrilateral and triangular elements. The capacity of CODE2 is restricted to not more than 450 nodal points and not more than 100 elements.

5.5.2 Data Storage in CODE2

The organization of data storage in CODE2 is essentially the same as that in CODE1. The main differences in storage are due to the following considerations:

1. The nodal point coordinate array must contain two values (x and y) for each node.

2. Only banded matrix storage is considered.

3. Only quadratic elements are used and the quadrature rules are fixed.

4. Boundary condition specification is much more complicated in two dimensions than in one. Boundary conditions may be applied to an arbitrary number of nodes (or element sides).

The data arrays for **CODE2**, as organized into common blocks, are described in Fig. 5.15.

Control Parameters

COMMON/RCON/TITLE(20), NNODE, NELEM, NMAT, NPOINT, NBC1, NBC2

     TITLE   = alphanumeric array containing problem title

     NNODE = number of nodal points in problem

     NELEM = number of elements in problem

     NMAT   = number of materials in problem

     NPOINT = number of point loads in problem

     NBC1   = number of nodes at which essential boundary conditions are prescribed

     NBC2   = number of element sides along which natural boundary conditions are prescribed

Nodal-Point Data

COMMON/CNODE/X(2,450), U(450)

     X = nodal point x- and y-coordinates

     U = nodal-point values of finite element approximation

Element Data

COMMON/CELEM/NE(100), MAT(100), NODES(9,100)

     NE   = number of nodes in element

     MAT   = material number of element

     NODES = node numbers of nodes in element

Material Data

COMMON/CMATL/PROP(3,5)

     PROP = material properties k, b, and f for up to five materials

FIGURE 5.15 *Data arrays for CODE2.*

Boundary-Condition Data

COMMON/CBC/NODBC1(70), VBC1(70), NELBC(40), NSIDE(40), VBC(2,40)
            NPT(10), VPT(10)

     NODBC1 = node numbers of nodes at which essential boundary
                conditions $u = \hat{u}$ are prescribed — see (4.2.18c)

     VBC1   = values of essential boundary conditions u

     NELBC   = element numbers of elements on which natural boundary
                conditions are prescribed — see (4.2.18d)

     NSIDE   = side number of element on which natural boundary
                conditions are prescribed

     VBC   = values of p and γ (or $\hat{\sigma}$) for natural boundary conditions
                — see (4.2.18d)

Matrix Storage

     COMMON/CMATRX/GK(450,21), GF(450)

     GK = global stiffness matrix **K**

     GF = global load vector **F**

Numerical Quadrature Data

     COMMON/CINT/XIQ(2,9), XIT(2,7), WQ(9), WT(7)

     XIQ = ξ, η coordinates of integration points for square master
            element

     XIT = ξ, η coordinates of integration points for triangular
            master element

     WQ = integration weights for square element

     WT = integration weights for triangle element

FIGURE 5.15   *(cont.)*

5.5.3   Preprocessor Routines

The preprocessing or data-generation phase of finite element codes grows rapidly in importance with the number of dimensions of the problem. In CODE2 we suggest generation schemes for nodal coordinates, element definition, and the prescription of boundary conditions that are minimal in complexity (and utility). Much more elaborate schemes are in common use.

In the preprocessor unit of CODE2, routines PREP, RCON, and RMAT are essentially the same as in CODE1—only obvious changes are required. We give here coding and description of input data formats for the generation of nodal coordinates in routine RNODE, element data in routine RELEM, and boundary-condition data in routine RBC.

Subroutine RNODE: This routine reads data descriptive of a sequence of nodal points that lie along a line in the x, y-plane and then generates the coordinates of these nodes. The first data record read in RNODE gives the value NREC, the number of such lines to be defined.

N1   = node number of first node along the line

NUM   = number of nodes along the line

INC   = increment in node number

X1,Y1   = x, y-coordinates of first node

XN,YN = x, y-coordinates of last node

The coding of RNODE follows:

```
        SUBROUTINE RNODE
        COMMON/CCON/TITLE(20), NNODE
        COMMON/CNODE/X(2,450)
        READ,NREC
        DO 20 IREC=1,NREC
        READ,N1,NUM,INC,X1,Y1,XN,YN
        NUM1=NUM-1
        XNUM=NUM1
        X(1,N1)=X1
        X(2,N1)=Y1
        IF (NUM.EQ.1) GO TO 20
        DX=(XN-X1)/XNUM
        DY=(YN-Y1)/XNUM
        DO 10 N=1,NUM1
        XN=N
        IN=N1+N*INC
        X(1,IN)=X1+XN*DX
10      X(2,IN)=Y1+XN*DY
20      CONTINUE
C.....PRINT NODAL POINT COORDINATES
        WRITE(6,101)
        DO 30 N=1,NNODE
30      WRITE(6,102) N,X(1,N),X(2,N)
        RETURN
101     FORMAT(35H NODE NO    X-COORDINATE    Y-COORDINATE    )
102     FORMAT(I7,2E16.4)
        END
```

Subroutine RELEM: This routine reads data that define a sequence of elements such that the corresponding nodes of each element in the sequence differs by a given increment from those of the preceding element in the sequence. The first data record read in RELEM gives the value of NREC, the number of such sequences to be defined. For each of the next **NREC** records, the values of the following variables are read.

NUM = number of elements in the sequence

INC = increment in nodal numbers from element to element in sequence

NEE = number of nodes in each element

MATE = material property number for the elements

NODE(I) = nodal point numbers of nodes in first element in the sequence

The coding of **RELEM** is

```
      SUBROUTINE RELEM
      COMMON/CCON/TITLE(20),NNODE,NELEM
      COMMON/CELEM/NE(100),MAT(100),NODES(9,100)
      DIMENSION NODE(9)
C.....READ ELEMENT DATA
      READ,NREC
      NEL=0
      DO 20 IREC=1,NREC
      READ,NUM,INC,NEE,MATE,(NODE(I),I=1,NEE)
      N1=NEL+1
      NEL=NEL+NUM
      IF(NEL.GT.100) GO TO 99
      DO 20 N=N1,NEL
      NINC=(N-1)*INC
      DO 10 M=1,NEE
10    NODES(M,N)=NODE(M)+NINC
      NE(N)=NEE
20    MAT(N)=MATE
C.....PRINT ELEMENT DEFINITIONS
      WRITE(6,101)
      DO 30 N=1,NELEM
      NEN=NE(N)
30    WRITE(6,102)N,NE(N),MAT(N),(NODES(I,N),I=1,NEN)
      RETURN
99    WRITE(6,103)
101   FORMAT(19H ELEM NO NE MAT,20X,12HNODE NUMBERS  //)
102   FORMAT(I5,I7,I6,5X,9I5)
103   FORMAT(37H ELEMENT NUMBER EXCEEDS MAXIMUM VALUE )
      STOP
      END
```

Subroutine RBC: This routine reads data records that describe point loads, essential boundary conditions on nodes, and natural boundary conditions on element sides. For each set of data, the first record read is the value of NREC—the number of subsequent records to be read for that data set. The subsequent data records for each set have the following format:

Point loads: one record per point load.

N = the node number of the node at which the point load is applied

V = the value of the point load

Essential boundary conditions $u = \hat{u}$: each record defines a sequence of nodes with the same boundary condition.

N1   = node number of first node in the sequence

NUM = number of nodes in sequence

INC = increment in node number

V   = value of $\hat{u}$ in the essential boundary condition

Natural boundary conditions $-k(\partial u/\partial n) = pu - \gamma$: each record defines a sequence of elements that have the natural boundary condition prescribed along one side.

N1   = element number of first element in the sequence

NUM = number of elements in the sequence

INC = increment in element numbers

NS   = side number of element along which boundary condition is prescribed

P   = value of the coefficient p in (4.2.18d)

V   = value of γ in (4.2.18d)

The coding of routine **RBC** is

```
        SUBROUTINE RBC
        COMMON/CCON/TITLE(20),NNODE,NELEM,NMAT,NPOINT,NBC1,
      .     NBC2
        COMMON/CBC/NODBC1(70),VBC1(70),NELBC(40),NSIDE(41),
      .     VBC2(2,40),NPT(10),VPT(10)
C.....READ POINT LOADS
        READ,NREC
        NPOINT=NREC
        IF(NREC.EQ.0) GO TO 20
        DO 10 I=1,NREC
        READ,N,V
        NPT(I)=N
10      VPT(I)=V
```

```
C ..... READ ESSENTIAL BOUNDARY CONDITION DATA
20         READ,NREC
           NBC1=0
           IF(NREC.EQ.0) GO TO 40
           DO 30 J=1,NREC
           READ,N1,NUM,INC,V
           DO 30 I=1,NUM
           NBC1=NBC1+1
           N=N1+(I-1)*INC
           NODBC1(NBC1)=N
30         VBC1(NBC1)=V
C ..... READ NATURAL BOUNDARY CONDITION DATA
40         READ,NREC
           NBC2=0
           IF(NREC.EQ.0) GO TO 60
           DO 50 J=1,NREC
           READ,N1,NUM,INC,NS,P,V
           DO 50 I=1,NUM
           NBC2=NBC2+1
           N=N1+(I-1)*INC
           NELBC(NBC2)=N
           NSIDE(NBC2)=NS
           VBC2(1,NBC2)=P
50         VBC2(2,NBC2)=V
C ..... PRINT BOUNDARY CONDITION DATA
           .
           .
           .

      (THE REMAINDER OF THIS ROUTINE IS LEFT AS AN EXERCISE)
           .
           .

           .
           END
```

EXERCISE

Programming Assignment 5.1:

Complete the coding of the main program and preprocessor for CODE2. Compile and correct *FORTRAN* errors in the coding. Prepare data sets for the example problem shown in Fig. 5.16. Run this data set and verify from the output that the preprocessor for CODE2 is operating correctly.

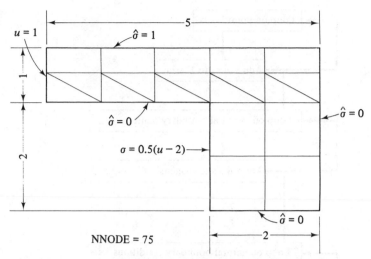

FIGURE 5.16 *Example problem for verification of preprocessor.*

5.5.4 Processor Routines

The computational details in the two-dimensional problem are more involved than in the one-dimensional case but the general scheme is similar. Routine **PROC** is unchanged. The changes in **FORMKF** account for different numbers of nodal coordinates, and so on, required in two dimensions. The assembly routine **ASSMB** and solution routines **TRIB** and **RHSB** from Section 3.8.1, p. 120, are used unchanged (except for dimensions in common blocks).

Routines **SHAPE** and **SETINT** are to be modified to incorporate the shape functions and integration rules described in Section 5.4. These modifications are discussed in the Exercises at the end of this section. The calculation of element stiffness matrices and load vectors is carried out in routines **ELEM** and **BCINT**. The area integral contributions to $\mathbf{k}^e$ and $\mathbf{f}^e$ are calculated in **ELEM** and the integrals over element boundaries leading to $\mathbf{P}^e$ and $\boldsymbol{\gamma}^e$ are calculated in **BCINT**. Subroutine **ELEM**, as in **CODE1**, is called from **FORMKF**, while the new routine **BCINT** is called from **APLYBC** (see Fig. 5.17). The calculations contain the essential features of finite element multi-dimensional coding and are given in complete form.

Subroutine ELEM: The definitions of the *FORTRAN* variables used in **ELEM** are given next, followed by a listing of the routine.

N   = number of nodal points N_e (and shape functions) in the element

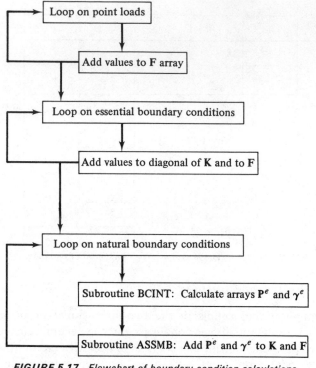

FIGURE 5.17 *Flowchart of boundary condition calculations.*

X(I,J)   = nodal-point coordinates, x_j and $y_j, j = 1, 2, \ldots, N_e$

   NOTE: X(1,J) $= x_j$ and   X(2,J) $= y_j$

NL   = number N_I of integration points

XI(J,L)   = coordinates (in the master-element coordinate system) ξ_l and $\eta_l, l = 1, 2, \ldots, N_l$

   NOTE: $\xi_l =$ XI(1,L)   and $\eta_l =$ XI(2,L)

W(L)   = integration weights $w_l, l = 1, 2, \ldots, N_l$

PSI(I)   = values of the shape functions $\hat{\psi}_i(\xi, \eta), i = 1, 2, \ldots, N_e$ at an integration point

DPSI(I,J)   = values of the shape-function gradients $\partial\hat{\psi}_i/\partial\xi$ and $\partial\hat{\psi}_i/\partial\eta$, $i = 1, 2, \ldots, N_e$, at an integration point

   NOTE: $\partial\hat{\psi}_i/\partial\xi =$ DPSI(I,1)   and $\partial\hat{\psi}_i/\partial\eta =$ DPSI(I,2)

DXDS(I,J)   = values the entries in the Jacobian matrix **J** of (5.2.2) at an integration point

   NOTE: $\partial x/\partial\xi =$ DXDS(1,1) $\partial x/\partial\eta =$ DXDS(1,2)
   $\partial y/\partial\xi =$ DXDS(2,1) $\partial y/\partial\eta =$ DXDS(2,2)

DSDX(I,J) = values of the entries of the inverse Jacobian matrix $\mathbf{J}^{-1}$ of (5.2.5) at an integration point

> *NOTE:* $\partial\xi/\partial x$ = DSDX(1,1) $\partial\xi/\partial y$ = DSDX(1,2)
> $\partial\eta/\partial x$ = DSDX(2,1) $\partial\eta/\partial y$ = DSDX(2,2)

XK,XB,XF = values of the coefficients in the differential equation (4.2.18a) for an element

EK   = element stiffness array

EF   = element load vector array

The transmission of data to subroutine ELEM follows the same pattern as in CODE1. The coding below follows directly from the list of steps in Section 5.3.2.

```
          SUBROUTINE ELEM(X,N,EK,EF,NL,XI,W,MAT)
          DIMENSION X(2,N),EK(N,N),EF(N),XI(2,NL),W(NL)
          DIMENSION DPSIX(9),DPSIY(9),DXDS(2,2),DSDX(2,2)
          DIMENSION PSI(9),DPSI(2,9)
C.....INITIALIZE ELEMENT ARRAYS
          DO 10 I=1,N
          EF(I)=0.0
          DO 10 J=1,N
10        EK(I,J)=0.0
          CALL GETMAT(XK,XB,XF,MAT)
C.....BEGIN INTEGRATION POINT LOOP
          DO 50 L=1,NL
          CALL SHAPE(XI(1,L),N,PSI,DPSI)
C.....CALCULATE DXDS ... EQUATION (5.3.6)
          DO 20 I=1,2
          DO 20 J=1,2
          DXDS(I,J)=0.0
          DO 20 K=1,N
20        DXDS(I,J)=DXDS(I,J)+DPSI(K,J)*X(I,K)
C.....CALCULATE DSDX ... EQUATION (5.2.7)
          DETJ=DXDS(1,1)*DXDS(2,2)-DXDS(1,2)*DXDS(2,1)
          IF(DETJ.LE.0. ) GO TO 99
          DSDX(1,1)=DXDS(2,2)/DETJ
          DSDX(2,2)=DXDS(1,1)/DETJ
          DSDX(1,2)=-DXDS(1,2)/DETJ
          DSDX(2,1)=-DXDS(2,1)/DETJ
C.....CALCULATE D(PSI)/DX ... EQUATION (5.3.5)
          DO 30 I=1,N
          DPSIX(I)=DPSI(I,1)*DSDX(1,1)+DPSI(I,2)*DSDX(2,1)
30        DPSIY(I)=DPSI(I,1)*DSDX(1,2)+DPSI(I,2)*DSDX(2,2)
```

```
C . . . . . ACCUMULATE INTEGRATION POINT VALUE OF INTEGRALS
          FAC=DETJ*W(L)
          DO 40 I=1,N
          EF(I)=EF(I)+XF*PSI(I)*FAC
          DO 40 J=I,N
   40     EK(I,J)=EK(I,J)+FAC*(XK*(DPSIX(I)*DPSIX(J)+DPSIY(I)*DPSIY(J))
        .    +XB*PSI(I)*PSI(J))
   50     CONTINUE
C . . . . . CALCULATE LOWER SYMMETRIC PART OF EK
          DO 60 I=1,N
          DO 60 J=1,I
   60     EK(I,J)=EK(J,I)
          RETURN
   99     WRITE(6,110) DETJ,X
  110     FORMAT(13H BAD JACOBIAN,E10.3/9E10.3/9E10.3)
          STOP
          END
```

Subroutine APLYBC: It is clear that the imposition of boundary conditions in two-dimensional problems is quite different from the one-dimensional case. In two dimensions, the specification of the location as well as the values of boundary conditions requires considerably more data.

The coding of boundary-condition calculations is given in complete detail. A flowchart of the operations performed by routine APLYBC is given in Fig. 5.17. Coding of routine APLYBC follows.

```
          SUBROUTINE APLYBC
          COMMON/CCON/TITLE(20),NODE,NELEM,NMAT,NPOINT,NBC1,NBC2
          COMMON/CBC/NODBC1(70),VBC1(70),NELBC(40),NSIDE(41)
        .    ,VBC2(2,40),NPT(10),VPT(10)
          COMMON/CELEM/NE(100),MAT(100),NODES(9,100)
          COMMON/CMATRX/GK(450,21),GF(450)
          COMMON/CNODE/X(2,450)
          DIMENSION NOD(3), XBC(2,3),PE(3,3),GAME(3)
C . . . . . APPLY POINT LOADS
          IF (NPOINT.EQ.0) GO TO 20
          DO 10 I=1,NPOINT
          N=NPT(I)
   10     GF(N)=GF(N)+VPT(I)
C . . . . . APPLY ESSENTIAL BOUNDARY CONDITIONS
   20     IF (NBC1.EQ.0) GO TO 40
          BIG=1.E30
          DO 30 I=1,NBC1
          N=NODBC1(I)
          GK(1,N)=BIG
   30     GF(N)=BIG*VBC1(I)
```

```
C . . . . . APPLY NATURAL BOUNDARY CONDITIONS
40      IF (NBC2.EQ.0) GO TO 70
        DO 60 I=1,NBC2
C . . . . . PICK OUT NODES ON SIDE OF ELEMENT
        NEL=NELBC(I)
        NS=NSIDE(I)
        NC=4
        IF(NE(NEL).EQ.6) NC=3
        NOD(1)=NS
        NOD(2)=NS+NC
        NOD(3)=NS+1
        IF(NS.EQ.NC) NOD(3)=1
C . . . . . PICK OUT NODAL COORDINATES
        DO 50 J=1,3
        NJ=NOD(J)
        NOD(J)=NODES(NJ,NEL)
        NJ=NOD(J)
        XBC(1,J)=X(1,NJ)
50      XBC(2,J)=X(2,NJ)
C . . . . . CALL BCINT TO CALCULATE BOUNDARY INTEGRALS PE AND
   .    GAME
        CALL BCINT(VBC2(1,I),VBC2(2,I),XBC,PE,GAME)
C . . . . . CALL ASSEMB TO ADD PE TO GK AND GAME TO GF
        CALL ASSMB(PE,GAME,3,NOD,GK,GF,450)
60      CONTINUE
70      RETURN
        END
```

EXERCISES

Programming Assignment 5.2:

1. Code a shape function routine SHAPE for the calculation of the values of the shape function $\hat{\psi}_i$ and their derivatives $\partial\hat{\psi}_i/\partial\xi$ and $\partial\hat{\psi}_i/\partial\eta$. The argument list should agree with the call to SHAPE from routine ELEM. Provide for the use of the 8-node and 9-node quadrilaterals of Fig. 5.9.

Code a short main program to test routine SHAPE by calling SHAPE for several values of element coordinates ξ and η and printing the results for comparison with hand calculation. The test program should also calculate and print check sums a, b, and c as in Programming Assignment 3.2, part 1.

2. Code a two-dimensional integration rule routine SETINT. Use the 3×3 quadrature rule of Fig. 5.10.

Code a short testing program to perform numerical integration over the two-dimensional domain Ω given by $-1 \leq x \leq 1$, $-1 \leq y \leq 1$. Use the quadrature data from SETINT to calculate approximations to the integrals

$$I_1 = \int_\Omega \cos \frac{\pi x}{2} \cos \frac{\pi y}{2} \, dx \, dy$$

$$I_2 = \int_\Omega (3x + 2x^2 y^2 + 7x^5 y^5) dx \, dy$$

3. Code a routine that, taking the place of routine PROC, utilizes data from PREP and performs finite element interpolation. For a given finite element mesh, the nodal point values of a given function g should be obtained by a call to subroutine G and these values used to construct the finite element interpolant g_h.

Verify, by numerical experiment, that an arbitrary arrangement of 9-node rectangular elements will interpolate exactly the function $g_1 : g_1(x, y) = 3x^2 y^2 + 2xy + 7$ while the interpolation error for the function $g_2 : g_2(x, y) = x^4 y^3$ depends on the element size.

4. Extend the capabilities of routine SHAPE of part 1 to include the quadratic triangular element of Fig. 5.13b. Notice that it is necessary to identify two of the area coordinates with ξ and η (as in (5.4.8)) so that the derivatives $\partial \hat{\psi}_i / \partial \xi$ and $\partial \hat{\psi}_i / \partial \eta$ become

$$\frac{\partial \hat{\psi}_i}{\partial \xi} = -\frac{\partial \hat{\psi}_i}{\partial \zeta_1} + \frac{\partial \hat{\psi}_i}{\partial \zeta_2} + \frac{\partial \hat{\psi}_i}{\partial \zeta_3}, \quad \frac{\partial \hat{\psi}_i}{\partial \eta} = -\frac{\partial \hat{\psi}_i}{\partial \zeta_1} + \frac{\partial \hat{\psi}_i}{\partial \zeta_2} + \frac{\partial \hat{\psi}_i}{\partial \zeta_3}$$

Extend routine SETINT of part 2 to include the cubic quadrature rule for triangular elements given in Fig. 5.14. Notice that the values of area coordinates given in Fig. 5.14 must be transformed to values of ξ and η (by means of (5.4.8)). Common block CINT should include separate arrays for the triangular and quadrilateral quadrature data.

5. Complete the coding of CODE2 through the processor unit. Add temporary print statements in FORMKF so that element and global arrays can be printed at various stages of the calculations.

6. Run a set of data for a small problem (say, two elements) through CODE2 with the temporary print statements operational. Check, by means of hand calculation, that typical entries in the element arrays are being calculated accurately. Verify, by observation of the arrays GK and GF, that APLYBC is operating correctly.

7. Devise a problem with a solution that is, say, bilinear in x and y. Run this problem and verify that the correct nodal values are calculated.

5.5.5 Postprocessor

The postprocessing functions included in CODE2 are much simpler than those of CODE1 : here we provide only for the printing of values of the finite element approximation u_h and the flux $\sigma_h = -k(\partial u_h / \partial x, \partial u_h / \partial y)$. A simple coding structure, consisting of a calling routine POST, a routine PRINTU to print nodal point values of u_h, and a routine FLUX to calculate and print values of σ_h will suffice. Routines POST and PRINTU require no further explanation.

The calculation of σ_h in FLUX follows the scheme in ELEM for the calculation of $\partial\hat{\psi}_i/\partial x$ and $\partial\hat{\psi}_i/\partial y$ but now used for the evaluation of $\partial u/\partial x$ and $\partial u/\partial y$. There is a significant difference in that the values of u_i are now known. Thus, $\partial u/\partial\xi$ and $\partial u/\partial\eta$ can be calculated directly as

$$\frac{\partial u}{\partial\xi} = \sum_{j=1}^{N_i} \frac{\partial\hat{\psi}_j}{\partial\xi} u_j \quad \text{and} \quad \frac{\partial u}{\partial\eta} = \sum_{j=1}^{N_i} \frac{\partial\hat{\psi}_j}{\partial\eta} u_j$$

Of course, the values of the shape-function gradients at the points at which the flux is to be calculated are obtained from calls to SHAPE.

The only remaining question is the location of the flux calculation points. For the quadratic elements of CODE2, the 2×2 Gauss points are often chosen for flux evaluation. The coding and debugging of the postprocessor as described above are left as an exercise.

EXERCISE

Programming Assignment 5.3:

1. Code the postprocessor.

2. Devise a verification problem with quadratic polynomial solution. Run the problem and report the results.

3. Prepare a user's guide describing the use of CODE2. Include scope and limitation of the code and a description of all input data and printed output.

6

EXTENSIONS

6.1 INTRODUCTION

We have now reached a point where most of the ideas essential to an introduction to finite element methods have been covered. There are, however, several remaining concepts not dealt with earlier which deserve some attention in a first course on the subject. These are natural extensions of previous results to:

1. Three-dimensional problems.
2. Fourth-order boundary-value problems.
3. Systems of differential equations (such as those encountered in problems in which the solution is vector-valued).
4. Time-dependent problems.

Our purpose here is to give a brief account of extensions of ideas discussed earlier to each of the four subjects listed above. We concentrate on only the formulative aspects of the method as it applies to these problems, leaving the discussion of details and limitations to later volumes.

6.2 THREE-DIMENSIONAL PROBLEMS

All of the ideas covered in Chapters 4 and 5 for two-dimensional problems are easily extended to three dimensions. Of course, certain computational features will require special attention. For example, the use of reasonable meshes for three-dimensional problems typically leads to very large systems of equations for which special solution techniques may be necessary.

As an example, let us consider a generalization of the boundary-value problems considered earlier to three dimensions. Let Ω denote a three-dimensional domain, such as that shown in Fig. 6.1. The boundary $\partial\Omega$ is divided into two parts: $\partial\Omega_1$ on which essential boundary conditions are prescribed and $\partial\Omega_2$ on which natural boundary conditions are prescribed. Note that the boundary $\partial\Omega$ is a surface defined by parametric equations of the form

$$x = x(s_1, s_2), \qquad y = y(s_1, s_2), \qquad z = z(s_1, s_2)$$

s_1 and s_2 being real parameters.

We wish to solve the following variational boundary-value problem: find a scalar-valued function $u = u(x, y, z) \in H^1(\Omega)$ such that $u = \hat{u}$ on $\partial\Omega_1$ and

$$\int_\Omega (k \, \nabla u \cdot \nabla v + buv) \, dx \, dy \, dz + \int_{\partial\Omega_2} puv \, ds = \int_\Omega fv \, dx \, dy \, dz + \int_{\partial\Omega_2} \gamma v \, ds$$

$$(6.2.1)$$

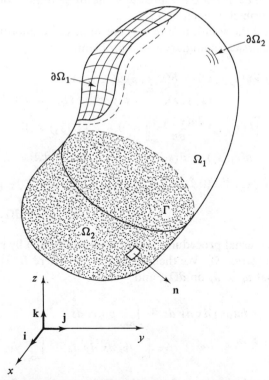

FIGURE 6.1 *Three-dimensional domain Ω containing an interface Γ between materials with different moduli.*

for all test functions $v \in H^1(\Omega)$ such that $v = 0$ on $\partial\Omega_1$. Here the coefficients $k = k(x, y, z)$, $b = b(x, y, z)$, the functions $p = p(s_1, s_2)$, $f = f(x, y, z)$, and $\gamma = \gamma(s_1, s_2)$ are given data in the problem and $\mathbf{V}u$ is the three-dimensional gradient,

$$\mathbf{V}u = \frac{\partial u}{\partial x}\mathbf{i} + \frac{\partial u}{\partial y}\mathbf{j} + \frac{\partial u}{\partial z}\mathbf{k}$$

$\mathbf{i}$, $\mathbf{j}$, and $\mathbf{k}$ being mutually orthogonal, unit basis vectors. The space $H^1(\Omega)$ of admissible functions now consists of those functions v such that

$$\int_\Omega \left[\left(\frac{\partial v}{\partial x}\right)^2 + \left(\frac{\partial v}{\partial y}\right)^2 + \left(\frac{\partial v}{\partial z}\right)^2 + v^2 \right] dx\, dy\, dz < \infty \qquad (6.2.2)$$

In (6.2.1), the modulus k and the coefficient b are assumed to be smooth functions, with $|k(x, y, z)| \geq k_0 = \text{constant} > 0$, except that they may undergo a simple jump discontinuity at a surface Γ defining the interface of two different materials, Ω_1 and Ω_2. Also, ds in (6.2.1) now denotes an element of surface area. The situation is precisely the three-dimensional version of the problem described in Section 4.2.

If the data and the solution of (6.2.1) are sufficiently smooth, u is also the solution of the classical boundary-value problem,

$$\left.\begin{aligned}
-\mathbf{V} \cdot [k(x, y, z)\, \mathbf{V}u(x, y, z)] + b(x, y, z)\, u(x, y, z) & \\
= f(x, y, z), \qquad (x, y, z) \in \Omega_i, \quad i = 1, 2 & \\
\left[\!\left[k(s_1, s_2) \frac{\partial u(s_1, s_2)}{\partial n} \right]\!\right] = 0, \qquad (s_1, s_2) \in \Gamma & \\
u(s_1, s_2) = \hat{u}(s_1, s_2), \qquad (s_1, s_2) \in \partial\Omega_1 & \\
k(s_1, s_2) \frac{\partial u(s_1, s_2)}{\partial n} + p(s_1, s_2)\, u(s_1, s_2) = \gamma(s_1, s_2), & \\
(s_1, s_2) \in \partial\Omega_2 &
\end{aligned}\right\} \qquad (6.2.3)$$

Following the usual procedure, we approximate (6.2.1) by replacing Ω by a suitable discretization Ω_h. We then seek $u_h \in H^h$, where H^h is a subspace of $H^1(\Omega_h)$, such that $u_h = \hat{u}_h$ on $\partial\Omega_{1h}$ and

$$\int_{\Omega_h} (k\, \mathbf{V}u_h \cdot \mathbf{V}v_h + b u_h v_h)\, dx\, dy\, dz + \int_{\partial\Omega_{2h}} p u_h v_h\, ds$$
$$= \int_{\Omega_h} f v_h\, dx\, dy\, dz + \int_{\partial\Omega_{2h}} \gamma v_h\, ds \qquad (6.2.4)$$

for all $v_h \in H^h$ such that $v_h = 0$ on $\partial\Omega_{1h}$. This equation, of course, reduces to the linear system

$$\sum_{j=1}^{N} K_{ij}u_j = F_i, \qquad i = 1, 2, \ldots, N \qquad (6.2.5)$$

where the stiffness matrix $[K_{ij}]$ and the load vector $\{F_i\}$ are calculated in the usual way and $u_1, u_2, \ldots, u_N$ are the nodal values of u_h. The calculation of the approximate flux and other features of the solution proceeds as in the two-dimensional case.

The principal features here that differ from our study of the one- and two-dimensional problems are the geometry of the finite elements and the nature of the shape functions ψ_i^e. Generally speaking, many of the same rules still apply:

1. We regard the mesh Ω_h as being described by a sequence of transformations from a fixed master element $\hat{\Omega}$ onto each finite element $\Omega_e, e = 1, 2, \ldots, E$.

2. The local shape functions ψ_i^e are images of polynomial shape functions $\hat{\psi}_i$ defined over $\hat{\Omega}$, and nodal points are located so that the final global basis functions ϕ_i $(i = 1, 2, \ldots, N)$ are continuous across interelement boundaries.

3. If $v_h(x, y, z) = \sum_{i=1}^{N} v_i\phi_i(x, y, z)$ is a typical test function in H^h, then v_h is in $H^1(\Omega_h)$ and, since

$$\phi_i(x_j, y_j, z_j) = \begin{cases} 1 & \text{if } i = j, \\ 0 & \text{if } i \neq j \end{cases} \qquad (6.2.6)$$

then　　　$v_i = v_h(x_i, y_i, z_i)$

Most three-dimensional elements are direct generalizations of elements encountered for two dimensions in Chapter 5, and our method of construction by mapping from master elements is formally unchanged. The master tetrahedron and master cube in Figs. 6.2 and 6.3 are the analogues of the master triangle and square in two dimensions. The actual elements Ω_e in the (x, y, z)-frame can be constructed by transformations from $\hat{\Omega}$ of the form

$$
\begin{aligned}
x(\xi, \eta, \zeta) &= \sum_{j=1}^{N_e} x_j\hat{\psi}_j(\xi, \eta, \zeta) \\
y(\xi, \eta, \zeta) &= \sum_{j=1}^{N_e} y_j\hat{\psi}_j(\xi, \eta, \zeta) \\
z(\xi, \eta, \zeta) &= \sum_{j=1}^{N_e} z_j\hat{\psi}_j(\xi, \eta, \zeta)
\end{aligned}
\right\} \qquad (6.2.7)
$$

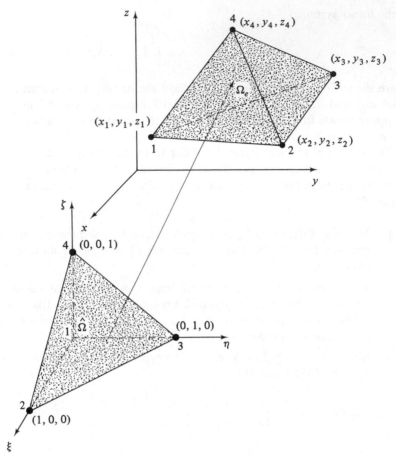

FIGURE 6.2 *Master tetrahedron $\hat{\Omega}$ and a corresponding element Ω_e.*

where $\hat{\psi}_j$ are the shape functions for polynomial interpolation on the master element.

A linear function in three dimensions involves four parameters (e.g., $f(\xi, \eta, \zeta) = a_1 + a_2\xi + a_3\eta + a_4\zeta$). Thus, the four linear shape functions $\{\hat{\psi}_j\} = \{\xi, \eta, \zeta, 1 - \xi - \eta - \zeta\}$ associated with the vertices of the master tetrahedron define a linear map to a tetrahedron Ω_e with planar faces. Similarly, trilinear shape functions $\hat{\psi}_j$, $j = 1, 2, \ldots, 8$, on the master cube define a map to a "brick" element Ω_e. Higher-degree shape functions can be formed as products of linear forms on the master element in essentially the same manner as in two dimensions. By using such shape functions in (6.2.7), three-dimensional elements with curved edges may be constructed.

The Jacobian $|\mathbf{J}|$ and coordinate derivatives $\partial \xi / \partial x, \ldots, \partial \zeta / \partial z$ can be

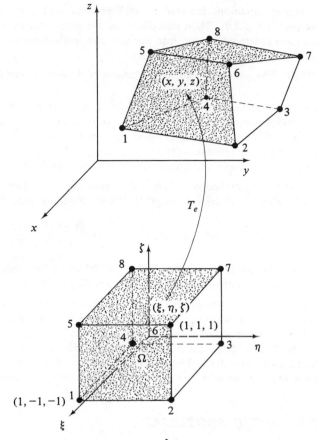

FIGURE 6.3 *Master cube $\hat{\Omega}$ and a corresponding brick element Ω_e.*

determined from (6.2.7) following the analysis of Section 5.2. The element calculations are performed with numerical quadrature on the master element $\hat{\Omega}$ as in Section 5.3.

EXERCISES

6.2.1   (a)   Furnish all steps in the derivation of the three-dimensional boundary-value problem (6.2.3), starting with a global conservation law and a linear constitutive equation as in Section 4.2.

   (b)   Using (6.2.3), develop the variational boundary-value problem (6.2.1) for sufficiently smooth u and v.

6.2.2 Derive explicit equations for element stiffness and load matrices for the discrete problem (6.2.3). Then describe how the global matrices **K** and **F** in (6.2.5) are generated, taking into account the nonhomogeneous natural boundary conditions.

[*REMARK:* This amounts to an easy extension of the ideas covered in Section 4.5.]

6.2.3 Consider the volume coordinates $\gamma_i = \hat{v}_i/\hat{V}$, $i = 1, 2, 3, 4$, for the master tetrahedron of volume $\hat{V}$ where $\hat{v}_i$ is the volume of the subtetrahedron formed by the triangular surface opposite vertex i and an interior point (ξ, η, ζ). Find the coordinate surfaces $\gamma_i = $ constant and use these coordinates to construct linear and quadratic shape functions $\hat{\psi}_j$ for $\hat{\Omega}$.

6.2.4 In the case of the tetrahedron, if the transformation T_e is linear, show that $|\mathbf{J}| = 6V_e$, where V_e is the volume of Ω_e. Verify the integration rule

$$\int_{\hat{\Omega}} \gamma_1^\alpha \gamma_2^\beta \gamma_3^\mu \gamma_4^\nu \, d\gamma_1 \, d\gamma_2 \, d\gamma_3 = \frac{\alpha! \, \beta! \, \mu! \, \nu!}{(\alpha + \beta + \mu + \nu + 3)!}$$

6.2.5 By means of tensor products of one-dimensional shape functions or otherwise, form representative trilinear and triquadratic shape functions on $\hat{\Omega}$.

6.2.6 The coding of the element stiffness and load matrix calculations for problem (6.2.3) is a direct extension of the coding for the two-dimensional case given in Section 5.5. Code a three-dimensional version of subroutine ELEM and the corresponding shape function and integration routines SHAPE and SETINT using trilinear shape functions and $2 \times 2 \times 2$ Gaussian quadrature. Calculate the stiffness of a unit cubical element with $k = 1$ and $b = 0$.

6.3 FOURTH-ORDER PROBLEMS

All of the boundary-value problems considered up to this point have been of second order. There are, however, many important physical situations described by boundary-value problems of fourth order. We now describe how the ideas and methods developed earlier can be extended to problems of this type.

6.3.1 Two-Point Boundary-Value Problems

Most of the basic ideas can be demonstrated for the one-dimensional case. Let us begin by considering a two-point problem in which the governing differential equation is of the form

$$a_0(x)\frac{d^4u(x)}{dx^4} + a_1(x)\frac{d^3u(x)}{dx^3} + a_2(x)\frac{d^2u(x)}{dx^2} + a_3(x)\frac{du(x)}{dx}$$
$$+ a_4(x)u(x) = f(x), \qquad 0 < x < l \tag{6.3.1}$$

This is a linear ordinary differential equation of fourth order with variable coefficients a_0, a_1, a_2, a_3, and a_4 defined on a domain Ω given as the interval $0 < x < l$. We assume that the coefficients a_i $(i = 0, \ldots, 4)$ are smooth bounded functions of x; in particular, a_0 is assumed to be such that $a_0(x) \geq$ constant > 0 for $0 \leq x \leq l$.

It is well known that when we "integrate" (6.3.1) we obtain a solution in which there appear exactly four arbitrary constants, so that four additional conditions must be imposed to determine the solution u uniquely. These additional conditions enter the problem as boundary conditions. We confine our attention to elliptic problems in which exactly half of these conditions are applied at each end of the domain: two conditions at $x = 0$ and two at $x = l$. Moreover, since (6.3.1) is of fourth order, the boundary conditions will involve no derivatives of u of order higher than 3. Thus, the most general linear boundary conditions for the types of problems we wish to consider are of the form

$$
\left.
\begin{array}{l}
\alpha_{01} u'''(0) + \beta_{01} u''(0) + \gamma_{01} u'(0) + \rho_{01} u(0) = \delta_{01} \\
\alpha_{02} u'''(0) + \beta_{02} u''(0) + \gamma_{02} u'(0) + \rho_{02} u(0) = \delta_{02} \\
\alpha_{l1} u'''(l) + \beta_{l1} u''(l) + \gamma_{l1} u'(l) + \rho_{l1} u(l) = \delta_{l1} \\
\alpha_{l2} u'''(l) + \beta_{l2} u''(l) + \gamma_{l2} u'(l) + \rho_{l2} u(l) = \delta_{l2}
\end{array}
\right\}
\tag{6.3.2}
$$

where $\alpha_{01}, \beta_{01}, \ldots, \delta_{l2}$ are given constants.

The constants $\alpha_{01}, \beta_{01}, \ldots, \rho_{l2}$ appearing as coefficients in (6.3.2) cannot be specified arbitrarily; the end conditions must be independent and they must be compatible with the differential equation in the sense that they lead to a valid integration-by-parts formula for our weighted-residual calculations. Assuming that these conditions are compatible with (6.3.1), the set of equations (6.3.1) and (6.3.2) define a fourth-order boundary-value problem in the real-valued function $u = u(x)$.

To produce a variational statement of problems (6.3.1) and (6.3.2), we follow the now-familiar procedure: multiply the residual ($r = a_0 u^{(iv)} + a_1 u'''$ $+ a_2 u'' + a_3 u' + a_4 u - f$) by a sufficiently smooth test function $v = v(x)$, integrate the product rv by parts, and equate the result to zero. This yields, for instance,

$$
\begin{aligned}
\int_0^l [(a_0 v)'' u'' &- (a_1 v)' u'' + a_2 u'' v + a_3 u' v + a_4 uv] \, dx \\
&= \int_0^l fv \, dx - a_0 vu''' \big|_0^l + [(a_0 v)' - a_1 v] u'' \big|_0^l
\end{aligned}
\tag{6.3.3}
$$

for all sufficiently smooth functions v. What we have done here is integrate the term $a_0 vu^{(iv)}$ twice by parts and the term $a_1 vu'''$ once by parts, so that no derivatives of u and v of higher order than 2 appear in the integrand. In this way we are guaranteed that in the final variational statement of our problem (to be given below), both u and v will share the same smoothness requirements. Alternative equivalent formulas could be obtained, for example, by integrating $a_2 vu''$ or $a_3 vu'$ by parts.

We next list three fundamentally important observations that lead us to a concrete definition of a variational statement of fourth-order problems of the type defined by (6.3.1) and (6.3.2):

1. First, if the boundary conditions are to be compatible with the governing differential equation (6.3.1), then it must be possible to incorporate them into integration-by-parts formulas such as (6.3.3). Otherwise, our problem is said to be "ill-posed" and there may exist no solution at all! In general, this requirement will be satisfied whenever it is possible to "solve" (6.3.2) for the endpoint derivatives of u of orders 2 and 3 in terms of the boundary data and endpoint derivatives and values of u. This ensures that (6.3.2) can be introduced on the right-hand side of (6.3.3) as natural boundary data.

Thus, the coefficients in (6.3.2) must be such that

$$\left.\begin{array}{l} \begin{vmatrix} \alpha_{01} & \beta_{01} \\ \alpha_{02} & \beta_{02} \end{vmatrix} = \alpha_{01}\beta_{02} - \beta_{01}\alpha_{02} \neq 0 \\[2em] \begin{vmatrix} \alpha_{l1} & \beta_{l1} \\ \alpha_{l2} & \beta_{l2} \end{vmatrix} = \alpha_{l1}\beta_{l2} - \beta_{l1}\alpha_{l2} \neq 0 \end{array}\right\} \tag{6.3.4}$$

in which case we solve (6.3.2) to obtain

$$\left.\begin{array}{l} u'''(0) = \mu_{01}u'(0) + \mu_{02}u(0) + \mu_0 \\ u''(0) = \lambda_{01}u'(0) + \lambda_{02}u(0) + \lambda_0 \\ u'''(l) = \mu_{l1}u'(l) + \mu_{l2}u(l) + \mu_l \\ u''(l) = \lambda_{l1}u'(l) + \lambda_{l2}u(l) + \lambda_l \end{array}\right\} \tag{6.3.5}$$

wherein the constants $\mu_{01}, \ldots, \lambda_l$ are determined by the coefficients in (6.3.2); for example,

$$\mu_{01} = \frac{\beta_{01}\gamma_{02} - \gamma_{01}\beta_{02}}{\alpha_{01}\beta_{02} - \beta_{01}\alpha_{02}}, \quad \mu_{02} = \frac{\beta_{01}\rho_{02} - \rho_{01}\beta_{02}}{\alpha_{01}\beta_{02} - \beta_{01}\alpha_{02}},$$

$$\ldots, \quad \lambda_l = \frac{\alpha_{l1}\delta_{l2} - \delta_{l1}\alpha_{l2}}{\alpha_{l1}\beta_{l2} - \beta_{l1}\alpha_{l2}}$$

2. Because of the selective integration by parts, the integral obtained in (6.3.3) involves products of derivatives of trial functions u and test functions v of order 2 and less. Thus, if we wish to identify a class of admissible func-

tions on which smoothness requirements are barely strong enough to make this integral well defined, it is sufficient to take u and v to be members of a class of functions, denoted H^2, whose derivatives of order 2 and less are square-integrable over Ω. In other words, a test function v will belong to H^2 if

$$\int_0^l [(v'')^2 + (v')^2 + v^2] \, dx < \infty \qquad (6.3.6)$$

3.   Suppose that u and v are in H^2. Then derivatives of u and v of third order and higher may not exist, and second derivatives may not exist at discrete points in Ω (e.g., v'' may suffer jumps at various points in the domain). It follows that if v is barely smooth enough to belong to H^2, it is impossible to impose conditions on derivatives of v of order 2 and higher (see Exericse 6.3.2). From this observation, there follows a fundamental property of two-point boundary-value problems of the type considered here: boundary conditions naturally fall into two distinct categories, those involving first derivatives and values of the solution (which can have meaning for functions in H^2) and those involving derivatives of second and third order, which must be enforced in some special way in the statement of the variational problem. Those boundary conditions involving values of u and its first derivative are *essential* boundary conditions; those involving derivatives of second and third order are *natural* boundary conditions. The essential boundary conditions enter the problem in the definition of the space of admissible functions, whereas the natural boundary conditions dictate the actual form of the variational equation.

Returning now to (6.3.3) and (6.3.5), we see that the variational statement for our model fourth-order boundary-value problem takes the following form: find $u \in H^2$ such that

$$\int_0^l [(a_0 v)''u'' - (a_1 v)'u'' + a_2 u''v + a_3 u'v + a_4 uv] \, dx$$

$$= \int_0^l fv \, dx - a_0(l)v(l)[\mu_{11}u'(l) + \mu_{12}u(l) + \mu_l]$$

$$+ a_0(0)v(0)[\mu_{01}u'(0) + \mu_{02}u(0) + \mu_0] \qquad (6.3.7)$$

$$+ [(a_0 v)'(l) - a_1(l)v(l)][\lambda_{11}u'(l) + \lambda_{12}u(l) + \lambda_l]$$

$$- [(a_0 v)'(0) - a_1(0)v(0)][\lambda_{01}u'(0) + \lambda_{02}u(0) + \lambda_0]$$

$$\text{for all } v \in H^2$$

We observe that whenever a solution to (6.3.7) is sufficiently smooth, it is also a solution of the problem defined by (6.3.1) and (6.3.2). Again note that the data in the natural boundary conditions enter in the statement of the variational problem and appear with f on the right-hand side of (6.3.7). If we have essential boundary conditions of the form

$$\left.\begin{array}{ll} \gamma_{01}u'(0) + \rho_{01}u(0) = \delta_{01}, & \gamma_{11}u'(l) + \rho_{11}u(l) = \delta_{11} \\ \gamma_{02}u'(0) + \rho_{02}u(0) = \delta_{02}, & \gamma_{12}u'(l) + \rho_{12}u(l) = \delta_{12} \end{array}\right\}$$ (6.3.8)

then the problem (6.3.7) reduces to one of finding $u \in H^2$ which satisfies (6.3.8) and

$$\int_0^l [(a_0 v)''u'' - (a_1 v)'u'' + a_2 u''v + a_3 u'v + a_4 uv]\, dx$$

$$= \int_0^l fv\, dx \qquad \text{for all } v \in H_0^2$$ (6.3.9)

wherein H_0^2 denotes the class of functions v satisfying (6.3.6) and which have the property that $v(0) = v'(0) = v(l) = v'(l) = 0$.

6.3.2 Finite Element Approximations

The construction of a finite element approximation of fourth-order problems, such as (6.3.7), is, as usual, based on Galerkin's method. Thus, we replace (6.3.7) by a finite-dimensional problem of the following form: find $u_h \in H^h$ such that

$$\left.\begin{array}{l} \displaystyle\int_0^l [(a_0 v_h)''u_h'' - (a_1 v_h)'u_h'' + a_2 u_h''v_h + a_3 u_h'v_h + a_4 u_h v_h]\, dx \\[2mm] \displaystyle= \int_0^l fv_h\, dx - a_0(l)v_h(l)[\mu_{11}u_h'(l) + \mu_{12}u_h(l) + \mu_l] \\[2mm] + a_0(0)v_h(0)[\mu_{01}u_h'(0) + \mu_{02}u_h(0) + \mu_0] \\[2mm] + [(a_0 v_h)'(l) - a_1(l)v_h(l)][\lambda_{11}u_h'(l) + \lambda_{12}u_h(l) + \lambda_l] \\[2mm] - [(a_0 v_h)'(0) - a_1(0)v_h(0)][\lambda_{01}u_h'(0) + \lambda_{02}u_h(0) + \lambda_0] \\[2mm] \hfill \text{for all } v_h \in H^h \end{array}\right\}$$ (6.3.10)

where H^h is an appropriate finite-dimensional subspace of H^2.

We outline the essential features of the method as follows:

1. The fact that the finite element space H^h of admissible functions for the approximate problem must be a subspace of H^2 is critical. This means that each test function v_h must be such that $(v_h'')^2$ is integrable over the interval Ω $(0 < x < l)$. What restriction does this put on our choices of basis functions? If a basis function ϕ_i is such that its derivative ϕ_i' has a discontinuity at a nodal point x_j, then ϕ_i'' behaves like a Dirac delta at x_j and, of course, $(\phi_i'')^2$ cannot be integrated over Ω. It follows that if each ϕ_i is to belong to H^2 (i.e., if each ϕ_i is to be such that $(\phi_i'')^2$ is integrable over Ω), then ϕ_i' cannot

have a discontinuity anywhere in Ω. In other words, *for problems of the form (6.3.7), the finite element basis functions ϕ_i must be such that their first derivatives are continuous throughout the domain of the approximate solution u_h.*

2. If the functions v_h on two adjacent elements, Ω_e and Ω_{e+1}, in a finite element mesh are to match at their common node in such a way that v_h' is continuous, it is necessary that both the values of v_h (i.e., v_h^e and v_h^{e+1}) and the values of its derivative v_h' ($v_h^{e'}$ and $v_h^{e+1'}$) coincide at this node. It follows that for fourth-order problems in one dimension, the test functions v_h assume the form*

$$v_h(x) = \sum_{i=1}^{N} v_i \phi_i^0(x) + \sum_{i=1}^{N} v_i' \phi_i^1(x)$$

where N is the number of nodes in the mesh and the basis functions ϕ_i^0, ϕ_i^1 have the properties

$$\left.\begin{array}{ll} \phi_i^0(x_j) = \begin{cases} 1 & \text{if } i = j, \\ 0 & \text{if } i \neq j \end{cases} & \dfrac{d\phi_i^0(x_j)}{dx} = 0 \\[3ex] \phi_i^1(x_j) = 0, & \dfrac{d\phi_i^1(x_j)}{dx} = \begin{cases} 1 & \text{if } i = j \\ 0 & \text{if } i \neq j \end{cases} \quad \text{for } 1 \leq i, j \leq N \end{array}\right\} \quad (6.3.11)$$

Clearly, $v_h(x_i) = v_i$ and $v_h'(x_i) = v_i'$, $i = 1, 2, \ldots, N$. Basis functions, such as those in (6.3.11), which interpolate derivatives as well as values at nodes, are called *Hermite* basis functions.

3. Hermite basis functions of the type in (6.3.11) can always be constructed by matching together element shape functions in the usual way. Let $\hat{\Omega}$ denote a master element, with coordinate ξ, $-1 \leq \xi \leq 1$. Note that the specification of u_h and u_h' at the endpoints of an element involves four conditions. Since a complete cubic in ξ contains exactly four parameters, we can construct cubic shape functions $\hat{\psi}_j^0$, $\hat{\psi}_j^1$ which exhibit the necessary properties:

$$\left.\begin{array}{llll} \hat{\psi}_1^0(-1) = 1, & \hat{\psi}_1^0(1) = 0, & \dfrac{d\hat{\psi}_1^0(-1)}{d\xi} = \dfrac{d\hat{\psi}_1^0(1)}{d\xi} = 0 \\[3ex] \hat{\psi}_2^0(-1) = 0, & \hat{\psi}_2^0(1) = 1, & \dfrac{d\hat{\psi}_2^0(-1)}{d\xi} = \dfrac{d\hat{\psi}_2^0(1)}{d\xi} = 0 \\[3ex] \hat{\psi}_1^1(-1) = \hat{\psi}_1^1(1) = 0, & \dfrac{d\hat{\psi}_1^1(-1)}{d\xi} = 1, & \dfrac{d\hat{\psi}_1^1(1)}{d\xi} = 0 \\[3ex] \hat{\psi}_2^1(-1) = \hat{\psi}_2^1(1) = 0, & \dfrac{d\hat{\psi}_2^1(-1)}{d\xi} = 0, & \dfrac{d\hat{\psi}_2^1(1)}{d\xi} = 1 \end{array}\right\} \quad (6.3.12)$$

---

* Basis functions that provide for the specification of even higher derivatives at the nodes (and which produce continuous derivatives of v_h of order 2 or higher) could also be used here (see Volume II).

These conditions uniquely define the local cubic *Hermite polynomial* shape functions,

$$\hat{\psi}_1^0(\xi) = \tfrac{1}{4}(2 - 3\xi + \xi^3), \qquad \hat{\psi}_2^0(\xi) = \tfrac{1}{4}(2 + 3\xi - \xi^3) \left.\begin{array}{c} \\ \\ \end{array}\right\} (6.3.13)$$
$$\hat{\psi}_1^1(\xi) = \tfrac{1}{4}(1 - \xi - \xi^2 + \xi^3), \qquad \hat{\psi}_2^1(\xi) = \tfrac{1}{4}(-1 - \xi + \xi^2 + \xi^3)$$

which are illustrated in Fig. 6.4.

The restriction v_h^e of a test function $v_h \in H^h$ to an element Ω_e between nodes $i - 1$ and i can be written

$$v_h^e(x) = \hat{v}(\xi(x)) \tag{6.3.14}$$

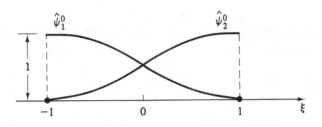

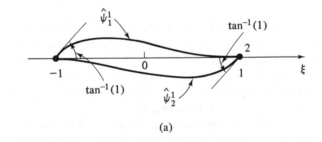

(a)

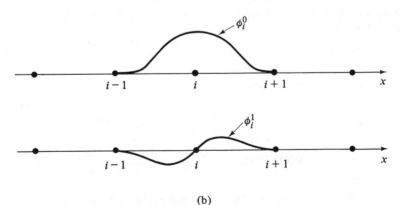

(b)

FIGURE 6.4 (a) *Hermite cubic shape functions on a master element and* (b) *corresponding global basis functions at a node i in the mesh.*

where

$$\xi(x) = \frac{2x - (x_{i-1} + x_i)}{x_i - x_{i-1}} \tag{6.3.15}$$

and $\hat{v}$ is the test function referred to the master element,

$$\hat{v}(\xi) = \sum_{j=1}^{2} v_j \hat{\psi}_j^0(\xi) + \sum_{j=1}^{2} \frac{h_e}{2} v_j' \hat{\psi}_j^1(\xi) \tag{6.3.16}$$

The element shape functions ψ_j^{0e}, ψ_j^{1e} are obtained from (6.3.13) and (6.3.15) as

$$\psi_j^{0e}(x) = \hat{\psi}_j^0(\xi(x)), \qquad \psi_j^{1e}(x) = \hat{\psi}_j^1(\xi(x)) \tag{6.3.17}$$

Combining corresponding shape functions ψ_j^{0e} and ψ_j^{0e+1} at node i determines the global basis function ϕ_i^0. Similarly, ψ_j^{1e} and ψ_j^{1e+1} at node i define the second global basis function ϕ_i^1 (Fig. 6.4).

Having selected the basis functions, the remainder of the analysis proceeds as in the study of second-order problems. The principal difference now is that the unknowns consist of not only the values $u_1, u_2, \ldots, u_N$ of u_h at the nodes, but also the nodal derivatives $u_1', u_2', \ldots, u_N'$. We leave additional details as exercises.

6.3.3 A Two-Dimensional Problem

A common two-dimensional, fourth-order boundary-value problem encountered in applications involves solving the biharmonic equation

$$\frac{\partial^4 u(x, y)}{\partial x^4} + 2\frac{\partial^4 u(x, y)}{\partial x^2 \partial y^2} + \frac{\partial^4 u(x, y)}{\partial y^4} = f(x, y), \qquad (x, y) \in \Omega \tag{6.3.18}$$

with boundary conditions of the type

$$u(s) = 0 \qquad \frac{\partial u(s)}{\partial n} = 0, \qquad s \in \partial\Omega \tag{6.3.19}$$

A variety of other boundary conditions could be imposed (see Exercise 6.3.3). Equations (6.3.18) and (6.3.19) describe, for example, the transverse deflection of a thin elastic plate clamped along its boundary $\partial\Omega$.

A variational statement of this problem is: find $u \in H_0^2(\Omega)$ such that

$$\int_\Omega \Delta u \, \Delta v \, dx \, dy = \int_\Omega fv \, dx \, dy \qquad \text{for all } v \in H_0^2(\Omega) \tag{6.3.20}$$

where $H_0^2(\Omega)$ is the class of all H^2 functions satisfying (6.3.19) and Δ is the

Laplacian operator,

$$\Delta u(x, y) = \frac{\partial^2 u(x, y)}{\partial x^2} + \frac{\partial^2 u(x, y)}{\partial y^2}$$

For more general boundary conditions, the left-hand side of (6.3.20) assumes a more complicated form (see Exercise 6.3.3).

Our objective here is to give some typical examples of finite elements appropriate for approximating (6.3.20). Since the test functions v_h for this case must be in H^2, it follows from earlier discussions that the basis functions ϕ_i must have continuous first partial derivatives across interelement boundaries (i.e., we must use C^1-functions for a basis). Since the normal derivative is to be continuous along an entire side rather than at a point, this condition is more demanding for elements in two dimensions than in the one-dimensional case. Two examples of elements that can be used for constructing such basis functions are:

1. **A C^1-Rectangle:** The product of a cubic in x by a cubic in y yields a collection of 16 monomials,

$$\begin{bmatrix} 1 \\ x \\ x^2 \\ x^3 \end{bmatrix} [1 \quad y \quad y^2 \quad y^3] = \begin{bmatrix} 1 & y & y^2 & y^3 \\ x & xy & xy^2 & xy^3 \\ x^2 & x^2y & x^2y^2 & x^2y^3 \\ x^3 & x^3y & x^3y^2 & x^3y^3 \end{bmatrix}$$

These provide a basis for the construction of a two-dimensional version of Hermite interpolation functions on a rectangular element in which values of the four quantities

$$u_h, \quad \frac{\partial u_h}{\partial x}, \quad \frac{\partial u_h}{\partial y}, \quad \frac{\partial^2 u_h}{\partial x \, \partial y}$$

are specified at each corner of the rectangle. If two such elements Ω_e and Ω_f have a common side in the mesh, then the specification that

$$\left\{ u_h^e, \frac{\partial u_h^e}{\partial x}, \frac{\partial u_h^e}{\partial y}, \frac{\partial^2 u_h^e}{\partial x \, \partial y} \right\} \quad \text{and} \quad \left\{ u_h^f, \frac{\partial u_h^f}{\partial x}, \frac{\partial u_h^f}{\partial y}, \frac{\partial^2 u_h^f}{\partial x \, \partial y} \right\}$$

have the same values at the common corner nodes ensures that u_h and $\partial u_h / \partial n$ (n being a normal to the common side) will be continuous on the interelement boundary (i.e., a C^1-finite element approximation is produced). Such an element is illustrated in Fig. 6.5a. In the figure, the following symbolism is used: the function value is indicated by a dot; a circle around a dot indicates that all first derivatives are

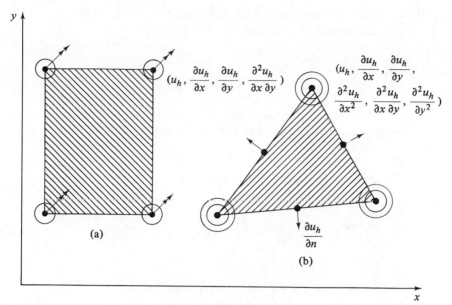

FIGURE 6.5 *Two examples of C¹-elements for fourth-order problems:* (a) *a bicubic rectangle and* (b) *a quintic triangle.*

prescribed and two circles imply second derivatives; a single-headed arrow indicates a normal derivative and a double-headed arrow denotes the mixed derivative (u_{xy}). The shape functions on the master square can be simply generated as a tensor product of the one-dimensional Hermite shape functions in (6.3.13).

2. **Quintic Triangle:** A complete quintic polynomial in x and y contains 21 terms. The specification of the six values

$$\left\{ u_h, \frac{\partial u_h}{\partial x}, \frac{\partial u_h}{\partial y}, \frac{\partial^2 u_h}{\partial x^2}, \frac{\partial^2 u_h}{\partial x \, \partial y}, \frac{\partial^2 u_h}{\partial y^2} \right\}$$

at each vertex and the value of the normal derivative

$$\left\{ \frac{\partial u_h}{\partial n} \right\}$$

at the midpoints of each side of a triangle supply the 21 degrees-of-freedom needed to determine a complete quintic polynomial. Moreover, the global basis functions ϕ_i produced by such quintic shape functions have continuous first partial derivatives across interelement boundaries, as required.

Other C^1-elements can be constructed and some additional examples are described in Volume II of this series.

EXERCISES

6.3.1 The following boundary-value problems are ill-posed or do not satisfy the assumptions described in the text. What is wrong with them?

(a) $\dfrac{1}{x-1}\dfrac{d^4u}{dx^4} + u = \sin x,\qquad 0 < x < 2$

$u(0) = u'(0) = 0,\qquad u(2) = u'(2) = 0$

(b) $\dfrac{d^4u}{dx^4} + u = 0,\qquad 0 < x < 1$

$\dfrac{d^4u(0)}{dx^4} = 1,\qquad \dfrac{d^3u(0)}{dx^3} = 0,\qquad u(1) = 0,\qquad \dfrac{du(1)}{dx} = 1$

(c) $\dfrac{d^4u}{dx^4} + x^2\dfrac{du}{dx} + u = x,\qquad 0 < x < 1$

$u(0) = u'(0) = u''(0) = 0,\qquad u(1) = u'(1) = 0$

6.3.2 The classical theory of bending of elastic beams is governed by the equation

$$\frac{d^2}{dx^2}\left[EI(x)\frac{d^2u(x)}{dx^2}\right] = f(x),\ 0 < x < l$$

in which u is the transverse deflection of the beam, EI is the bending stiffness, and f is the load per unit length. In addition to the deflection u, quantities of interest include the slope $\theta(x) = du(x)/dx$, the bending moment $M(x) = EI(x)[d^2u(x)/dx^2]$, and the shear $V(x) = (d/dx)[EI(x)d^2u(x)/dx^2]$.

(a) What restrictions are placed on the specification of boundary conditions in terms of u, θ, M, and V by (6.3.4)?

(b) Write the boundary terms of (6.3.7) in terms of u, θ, M, and V at $x = 0$ and $x = l$.

(c) Specialize the results of part (b) to the following cases:

(i) $u(0) = \theta(0) = 0,\qquad V(l) = V_l,\qquad M(l) = M_l$

(ii) $u(0) = M(0) = 0,\qquad \theta(l) = V(l) = 0$

(d) For $EI(x) =$ constant, use the shape functions of (6.3.13) in the variational statement (6.3.3) to calculate the element stiffness matrix $\mathbf{k}^e$ for a beam element of length h. Use the following ordering of the degrees-of-freedom for the element:

$$(\mathbf{u}^e)^T = [u_1^e, u_1^{e'}, u_2^e, u_2^{e'}]$$

(e) Calculate the element load vector $\mathbf{f}^e$ for the case of a uniformly distributed load ($f(x) =$ constant) on the element of part (d).

(f) Write the assembled stiffness matrix $\mathbf{K}$ and load vector $\mathbf{F}$ for a mesh of two equal elements for the problem shown below. Modify $\mathbf{K}$ and $\mathbf{F}$ to include the boundary conditions and solve the resulting system of equations. Sketch the solution $u_h = u_h(x)$.

[*NOTE:* The exact solution u of this boundary-value problem has a maximum value $u_{\max} = 5fl^4/384EI$ at $x = l/2$.]

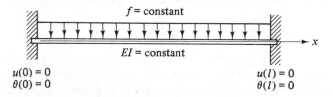

$$u(0) = 0 \qquad\qquad\qquad\qquad u(l) = 0$$
$$\theta(0) = 0 \qquad\qquad\qquad\qquad \theta(l) = 0$$

(g) Rework part (f) with the distributed load replaced by a concentrated load at $x = \frac{l}{2}$.

(h) In the case of natural boundary conditions of the form

$$EIu'''(0) = V_0 \qquad EIu'''(l) = V_l$$
$$EIu''(0) = M_0 \qquad EIu''(l) = M_l$$

the end shears and moments V_0, M_0, V_l, and M_l cannot be specified arbitrarily; they must be in equilibrium (compatible) with the applied load f. If u is any solution of the variational problem in this case, show that $u = c_1 + c_2 x$ is also a solution, c_1 and c_2 being arbitrary constants. From this fact, derive a pair of compatibility conditions on V_0, M_0, V_l, M_l, and $f = f(x)$ that must be satisfied if this problem is to have a solution.

6.3.3 Consider the biharmonic equation defined in (6.3.18). Suppose that Ω is a unit square with sides parallel to the x- and y-axis.

(a) By multiplying both sides of this equation by a smooth test function $v = v(x, y)$ and integrating twice by parts, show that (6.3.20) is an appropriate variational statement of this problem for the particular choice of boundary conditions (6.3.19).

(b) Using the results of part (a), derive an appropriate variational statement of (6.3.18) for cases in which, instead of (6.3.19), we have the following boundary conditions:

$$\left.\begin{array}{l} \dfrac{\partial^2 u}{\partial x^2} = M(y) \\[2mm] u = 0 \end{array}\right\} \quad \text{along } x = 0, \quad x = 1$$

$$\left.\begin{array}{l} \dfrac{\partial^2 u}{\partial y^2} = 0 \\[2mm] u = 0 \end{array}\right\} \quad \text{along } y = 0, \quad y = 1$$

6.3.4 A complete cubic in two dimensions involves 10 terms. Consider a triangle on which u_h is specified at the vertices and the centroid and $\partial u_h/\partial x$ and $\partial u_h/\partial y$ are prescribed at the vertices. Is this an appropriate element for fourth-order problems? Why?

6.4 SYSTEMS OF DIFFERENTIAL EQUATIONS

In many physical situations, we encounter boundary-value problems in which the primary dependent variable (the state variable) is a vector-valued function of position. Then the boundary-value problem involves a system of differential equations in the components of the unknown vector field **u**. For example, **u** may be the vector field of displacements for elasticity problems or velocities in flow problems; in transport processes the components of **u** may be the concentrations of different chemicals or species.

The variational formulation and finite element analysis for systems of differential equations is analogous to that employed throughout this text for scalar equations. We introduce the ideas first for a system comprised of a pair of ordinary differential equations and Dirichlet data. Following this, the important example of plane-stress problems in elasticity is considered.

6.4.1 A One-Dimensional System

As an illustrative example, let us consider the one-dimensional system of second-order equations,

$$\left.\begin{array}{l} -u''(x) + q(x)u(x) + r(x)\bar{u}(x) = f(x) \\ -\bar{u}''(x) + s(x)\bar{u}(x) + t(x)u(x) = g(x) \end{array}\right\} \quad 0 < x < 1 \qquad (6.4.1)$$

for the pair of functions $u = u(x)$ and $\bar{u} = \bar{u}(x)$ in $0 < x < 1$. Since each equation is of second order, two boundary conditions are required to specify each of the solution components u and $\bar{u}$ uniquely. For convenience, we assume homogeneous Dirichlet data at the ends,

$$u(0) = u(1) = \bar{u}(0) = \bar{u}(1) = 0 \qquad (6.4.2)$$

The data include the prescribed functions f, g, q, r, s, and t, which are assumed to be bounded and sufficiently smooth to ensure subsequent variational integrals are well defined and the problem "well posed."

The variational statement may be developed in a manner similar to that employed for the scalar two-point problem of Chapters 1 and 2. Letting v and $\bar{v}$ be test functions, we form the weighted-residual statement for each individual equation in (6.4.1). For u, $\bar{u}$, v, and $\bar{v}$ sufficiently smooth, we have

$$\left. \begin{array}{l} \int_0^1 (-u'' + qu + r\bar{u})v \, dx = \int_0^1 fv \, dx \\[2mm] \int_0^1 (-\bar{u}'' + s\bar{u} + tu)\bar{v} \, dx = \int_0^1 g\bar{v} \, dx \end{array} \right\} \qquad (6.4.3)$$

Integrating by parts and setting $v = \bar{v} = 0$ at the ends $x = 0$ and $x = 1$, we obtain the desired variational boundary-value problem: find $u \in H_0^1$ and $\bar{u} \in H_0^1$ such that

$$\left. \begin{array}{l} \int_0^1 (u'v' + quv + r\bar{u}v) \, dx = \int_0^1 fv \, dx \\[2mm] \int_0^1 (\bar{u}'\bar{v}' + s\bar{u}\bar{v} + tu\bar{v}) \, dx = \int_0^1 g\bar{v} \, dx \end{array} \right\} \qquad (6.4.4)$$

for all test functions $v \in H_0^1$ and $\bar{v} \in H_0^1$.

Each of the integral statements in (6.4.4) closely resembles that of our standard two-point problem in Chapters 1 through 3. In fact, if $r = t = 0$, the equations uncouple, forming a pair of distinct two-point problems. The details of the finite element derivations for (6.4.4) are similar to those encountered earlier and we shall only summarize the salient points here.

Since u, $\bar{u}$, v, and $\bar{v}$ are in H^1, a set of piecewise-polynomial basis functions $\{\phi_i\}$, $i = 1, 2, \ldots, N$, continuous on $0 \leq x \leq 1$, may be used to define the finite element space H_0^h. Here $i = 1, 2, \ldots, N$ denote the interior nodes in view of the fact that $u(0) = u(1) = 0$. The representations of functions v_h and $\bar{v}_h$ in H_0^h are of the form

$$v_h(x) = \sum_{j=1}^N v_j\phi_j(x) \qquad \text{and} \qquad \bar{v}_h(x) = \sum_{j=1}^N \bar{v}_j\phi_j(x) \qquad (6.4.5)$$

The finite element problem is to find $u_h \in H_0^h$ and $\bar{u}_h \in H_0^h$ such that

$$\left. \begin{array}{ll} \int_0^1 (u_h'v_h' + qu_hv_h + r\bar{u}_hv_h) \, dx = \int_0^1 fv_h \, dx & \text{for all } v_h \in H_0^h \\[2mm] \int_0^1 (\bar{u}_h'\bar{v}_h' + s\bar{u}_h\bar{v}_h + tu_h\bar{v}_h) \, dx = \int_0^1 g\bar{v}_h \, dx & \text{for all } \bar{v}_h \in H_0^h \end{array} \right\} \qquad (6.4.6)$$

Introducing representations of the form (6.4.5) into (6.4.6) leads to the finite element system

$$\left. \begin{array}{l} \sum_{j=1}^N \left\{ \int_0^1 [(\phi_i'\phi_j')u_j + (q\phi_i\phi_j)u_j + (r\phi_i\phi_j)\bar{u}_j] \, dx \right\} = \int_0^1 f\phi_i \, dx \\[3mm] \sum_{j=1}^N \left\{ \int_0^1 [(\phi_i'\phi_j')\bar{u}_j + (s\phi_i\phi_j)\bar{u}_j + (t\phi_i\phi_j)u_j] \, dx \right\} = \int_0^1 g\phi_i \, dx \\[3mm] \hspace{4cm} \text{for } i = 1, 2, \ldots, N \end{array} \right\} \qquad (6.4.7)$$

The element matrix contributions to the system have the same form as those in (6.4.7), with $\phi_i(x)$, $\phi_j(x)$ replaced by element shape functions $\psi_i^e(x)$, $\psi_j^e(x)$ throughout. Thus, the element calculations can be carried out and the system (6.4.7) assembled sequentially element-by-element in the usual manner. Both the number of equations and the bandwidths are increased over those of the scalar problem (see Exercise 6.4.3).

The method outlined above can be easily extended to more general forms of equations than (6.4.1), to more components, and to equations of higher order. Of greater interest is the extension to systems of partial differential equations such as those occurring in solid and fluid mechanics problems.

6.4.2   Plane-Stress Problems

We shall now outline the finite element analysis of a model problem in plane elasticity. Most of the ideas and technical details of the procedure we describe are readily extended to problems of fluid flow.

In elasticity, the state variable $\mathbf{u}$ is the displacement vector and the flux $\boldsymbol{\sigma}$ is the stress tensor. As our example, we study the plane-stress problem for an isotropic linearly elastic body. The governing equations can be written in concise form by introducing the matrices

$$
\boldsymbol{\sigma} = \begin{bmatrix} \sigma_{11} \\ \sigma_{22} \\ \sigma_{12} \end{bmatrix}, \quad
\boldsymbol{\epsilon} = \begin{bmatrix} \epsilon_{11} \\ \epsilon_{22} \\ \epsilon_{12} \end{bmatrix}, \quad
\mathbf{D} = \begin{bmatrix} \dfrac{\partial}{\partial x} & 0 \\ 0 & \dfrac{\partial}{\partial y} \\ \dfrac{\partial}{\partial y} & \dfrac{\partial}{\partial x} \end{bmatrix}, \quad
\mathbf{E} = \frac{E}{1 - \nu^2} \begin{bmatrix} 1 & \nu & 0 \\ \nu & 1 & 0 \\ 0 & 0 & \dfrac{1 - \nu}{2} \end{bmatrix},
$$

$$
\tag{6.4.8}
$$

$$
\mathbf{N} = \begin{bmatrix} n_x & 0 & n_y \\ 0 & n_y & n_x \end{bmatrix}, \quad
\mathbf{u} = \begin{bmatrix} u_x \\ u_y \end{bmatrix}, \quad
\mathbf{f} = \begin{bmatrix} f_x \\ f_y \end{bmatrix}, \quad
\hat{\boldsymbol{\sigma}} = \begin{bmatrix} \hat{\sigma}_x \\ \hat{\sigma}_y \end{bmatrix}
$$

Here σ_{ij} are the components of stress ($\sigma_{ij} = \sigma_{ji}$) and ϵ_{ij} are the components of strain ($\epsilon_{ij} = \epsilon_{ji}$); E and ν are Young's modulus of elasticity and Poisson's ratio, respectively, both constants characterizing the material; $\mathbf{D}$ is a matrix operator; n_x and n_y are the components of a unit vector normal to the boundary $\partial\Omega$ of a two-dimensional domain Ω; $\mathbf{u}$ is the displacement vector; $\mathbf{f}$ is a vector of components of internal sources representing the body force per unit area; and $\hat{\boldsymbol{\sigma}}$ is a vector representing the surface tractions applied to a portion $\partial\Omega_2$ of the boundary.

The conservation law for this problem, which expresses the principle of conservation of linear momentum for the body Ω (static equilibrium in this case), now assumes the form

$$\int_{\omega} (\mathbf{D}^T\boldsymbol{\sigma} + \mathbf{f})\, dx\, dy = \mathbf{0} \tag{6.4.9}$$

for an arbitrary portion ω of the body. If $\boldsymbol{\sigma}$ and $\mathbf{f}$ are sufficiently smooth, this leads to the system of partial differential equations of equilibrium for the body:

$$\mathbf{D}^T\boldsymbol{\sigma}(x, y) + \mathbf{f}(x, y) = \mathbf{0}, \qquad (x, y) \in \Omega \tag{6.4.10}$$

The constitutive equation characterizing the material as linearly elastic, homogeneous, and isotropic is simply

$$\boldsymbol{\sigma} = \mathbf{E}\boldsymbol{\epsilon} \tag{6.4.11}$$

where the strains $\boldsymbol{\epsilon}$ are related to the state variable $\mathbf{u}$ according to the strain-displacement relations,

$$\boldsymbol{\epsilon} = \mathbf{Du} \tag{6.4.12}$$

Introducing (6.4.11) and (6.4.12) into (6.4.10) gives the governing system of partial differential equations in terms of the displacement vector $\mathbf{u}$:

$$\mathbf{D}^T\mathbf{EDu}(x, y) + \mathbf{f}(x, y) = \mathbf{0}, \qquad (x, y) \in \Omega \tag{6.4.13}$$

To these we must add boundary conditions. The conservation principle, applied to the portion $\partial\Omega_2$ of $\partial\Omega$, yields the condition

$$\mathbf{N}\boldsymbol{\sigma} = \hat{\boldsymbol{\sigma}} \quad \text{or} \quad \mathbf{NEDu}(s) = \hat{\boldsymbol{\sigma}}(s), \qquad s \in \partial\Omega_2 \tag{6.4.14}$$

whereas the specification of the displacement vector $\mathbf{u}$ on the complementary portion $\partial\Omega_1$ of the boundary is characterized by the condition

$$\mathbf{u}(s) = \hat{\mathbf{u}}(s) = \begin{bmatrix} \hat{u}_x(s) \\ \hat{u}_y(s) \end{bmatrix}, \qquad s \in \partial\Omega_1 \tag{6.4.15}$$

Equations (6.4.13), (6.4.14), and (6.4.15) define a general two-dimensional boundary-value problem in linear elasticity, describing the problem of plane stress of an isotropic, linearly elastic body.

The variational statement of this problem is obtained in the usual way. If $\mathbf{v}$ is an arbitrary admissible displacement vector, then we wish to find $\mathbf{u}$ such that $u_x, u_y \in H^1(\Omega)$, $\mathbf{u} = \hat{\mathbf{u}}$ on $\partial\Omega_1$, and

$$\int_{\Omega} (\mathbf{Dv})^T\mathbf{ED}\,\mathbf{u}\, dx\, dy = \int_{\Omega} \mathbf{v}^T\mathbf{f}\, dx\, dy + \int_{\partial\Omega_2} \mathbf{v}^T\hat{\boldsymbol{\sigma}}\, ds \tag{6.4.16}$$
$$\text{for all } \mathbf{v} \in \hat{\mathbf{H}}^1(\Omega)$$

where $(\cdot)^T$ denotes transposition and $\hat{\mathbf{H}}^1(\Omega)$ is the space of vector-valued functions $\mathbf{v}$ with components v_i in $H^1(\Omega)$ such that $\mathbf{v} = \mathbf{0}$ on $\partial\Omega_1$. Equation (6.4.16) is known as the *principle of virtual work*; it represents a complete statement of the problem of equilibrium in elasticity in terms of the displacements.

The finite element approximation of (6.4.16) is obtained in the usual way except that now we must construct approximations of each component of $\mathbf{u}$. For a typical element Ω_e such as that shown in Fig. 6.6, we have

$$\int_{\Omega_e} (\mathbf{D}\mathbf{v}_h^e)^T \mathbf{E} \mathbf{D} \mathbf{u}_h^e \, dx \, dy = \int_{\Omega_e} \mathbf{v}_h^{eT} \mathbf{f} \, dx \, dy + \int_{\partial\Omega_e} \mathbf{v}_h^{eT} (\mathbf{N}\boldsymbol{\sigma}) \, ds \qquad (6.4.17)$$

where

$$\mathbf{u}_h^e(x, y) = \boldsymbol{\psi}^e(x, y)\mathbf{u}^e \qquad (6.4.18)$$

and

$$\boldsymbol{\psi}^e(x, y) = \begin{bmatrix} \psi_1^e(x, y) & 0 & \psi_2^e(x, y) & 0 & \psi_3^e(x, y) & 0 \\ 0 & \psi_1^e(x, y) & 0 & \psi_2^e(x, y) & 0 & \psi_3^e(x, y) \end{bmatrix}$$

$$(6.4.19)$$

$$(\mathbf{u}^e)^T = \{u_{x1}^e, u_{y1}^e, u_{x2}^e, u_{y2}^e, u_{x3}^e, u_{y3}^e\} \qquad (6.4.20)$$

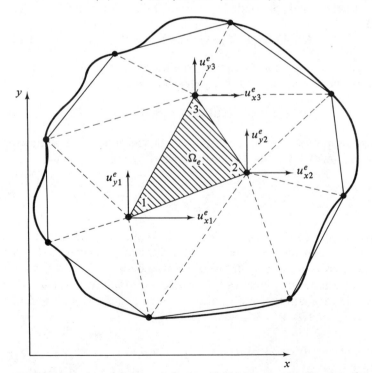

FIGURE 6.6 *A typical element Ω_e in a finite element approximation of a problem in plane elasticity.*

Here ψ_i^e are the usual element shape functions for element Ω_e and are no different from those we used earlier in scalar-valued problems. For example, in the case of the element shown in Fig. 6.6, these are the linear shape functions for a triangle described in Section 4.4. The equations governing a typical element are

$$\mathbf{k}^e \mathbf{u}^e = \mathbf{f}^e + \hat{\boldsymbol{\sigma}}^e \tag{6.4.21}$$

where

$$
\left.
\begin{aligned}
\mathbf{k}^e &= \int_{\Omega_e} (\mathbf{D}\boldsymbol{\psi}^e)^T \mathbf{E}\mathbf{D}\boldsymbol{\psi}^e \, dx \, dy \\[4pt]
\mathbf{f}^e &- \int_{\Omega_e} (\boldsymbol{\psi}^e)^T \mathbf{f} \, dx \, dy \\[4pt]
\hat{\boldsymbol{\sigma}}^e &= \oint_{\partial\Omega_e} (\boldsymbol{\psi}^e)^T \mathbf{N}\boldsymbol{\sigma} \, ds
\end{aligned}
\right\} \tag{6.4.22}
$$

For the triangular element in Fig. 6.6, $\mathbf{k}^e$ is a 6×6 matrix and $\mathbf{f}^e$ and $\boldsymbol{\sigma}^e$ are 6×1 vectors. The assembly of the global stiffness matrix and the application of boundary conditions now follow the procedure described in detail in Chapters 2 through 5.

EXERCISES

6.4.1 Write the fourth-order problem $u^{(iv)} = f$ as a system $u'' = \bar{u}$, $\bar{u}'' = f$, and develop a finite element formulation for the reduced problem. Discuss the merits of this approach in higher dimensions.

6.4.2 Use a mesh of four elements and compute solutions to (6.4.1) and (6.4.2) for $q = s = 0$, $r = 1$, $t = 2$, and $f = g = 1$. If $r = t = 1$ instead, then the system (6.4.1) can be simplified to a single equation. Show that this is also true of the finite element equations (6.4.7) in this case.

6.4.3 For the case where q, r, s, and t are constants, show that the element contribution for piecewise-linear approximation in (6.4.7) is

$$
\mathbf{k}^e \mathbf{u}^e = \frac{1}{h_e}
\begin{bmatrix}
1 & 0 & -1 & 0 \\
0 & 1 & 0 & -1 \\
-1 & 0 & 1 & 0 \\
0 & -1 & 0 & 1
\end{bmatrix}
\begin{bmatrix}
u_1 \\ \bar{u}_1 \\ u_2 \\ \bar{u}_2
\end{bmatrix}
$$

$$
+ \frac{h_e}{6}
\begin{bmatrix}
2q & 2r & q & r \\
2t & 2s & t & s \\
q & r & 2q & 2r \\
t & s & 2t & 2s
\end{bmatrix}
\begin{bmatrix}
u_1 \\ \bar{u}_1 \\ u_2 \\ \bar{u}_2
\end{bmatrix}
$$

6.4.4 Consider the system

$$\Delta u + \mathbf{a} \cdot \nabla u + bu + c\bar{u} = g$$

$$\Delta \bar{u} + \mathbf{d} \cdot \nabla \bar{u} + e\bar{u} + \acute{f}u = h$$

where Δ is the Laplacian, $\Delta = \nabla \cdot \nabla$, and $\mathbf{a}, b, \ldots, h$ are given smooth functions of x and y. Develop an integral statement of the system and a finite element formulation for a bounded two-dimensional domain Ω with Dirichlet data on $\partial\Omega$.

6.4.5 (a) Write out in detail the system of equations (6.4.10)–(6.4.13) for problems in plane elasticity.

(b) Derive the principle of virtual work, equation (6.4.16), writing out all terms in scalar form. Then show that if the quantities appearing in this equation are sufficiently smooth, (6.4.16) implies (6.4.13)–(6.4.15).

(c) Develop the final global equations describing the two-element approximation of the plane-stress problem indicated below in which Young's modulus $E = E_0$ and Poisson's ratio $v = \frac{1}{4}$.

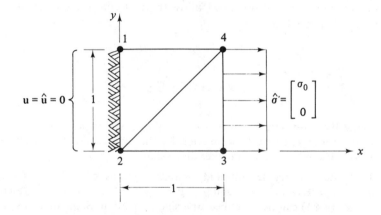

6.5 TIME-DEPENDENT PROBLEMS

Let us now turn to cases in which the dependent variable u is a function of both position and time. For simplicity, we confine our attention to problems involving only one spatial dimension.

In physical problems of the type considered in Chapters 2 and 4, the introduction of the dependence of the state variable u on time leads to conservation laws of the form

$$\frac{\partial}{\partial x}\sigma(x, t) - f(x, t) + \frac{\partial}{\partial t}\Gamma(x, t) = 0, \qquad 0 < x < l, \quad t > 0 \qquad (6.5.1)$$

where σ is the flux, f is the density of distributed internal sources, and Γ is the quantity being conserved in the physical process. For example, if energy is conserved $\partial \Gamma / \partial t$ might represent the time rate of change of entropy per unit length per degree absolute temperature, in which case it is related to the state variable u (the temperature) by

$$\frac{\partial \Gamma(x, t)}{\partial t} = c(x, t) \frac{\partial u(x, t)}{\partial t} \tag{6.5.2}$$

where $c(x, t)$ is a material property known as the specific heat. Alternatively, if (6.5.1) expresses the conservation of linear momentum in a deformable body and u is the displacement field, then Γ is the momentum at x at time t:

$$\Gamma(x, t) = \rho(x) \frac{\partial u(x, t)}{\partial t}; \qquad \frac{\partial \Gamma(x, t)}{\partial t} = \rho(x) \frac{\partial^2 u(x, t)}{\partial t^2} \tag{6.5.3}$$

where $\rho(x)$ is the mass density at point x.

As before, the flux is assumed to be given in terms of the spatial rate of change of u by a linear constitutive equation of the form

$$\sigma(x, t) = -k(x) \frac{\partial u(x, t)}{\partial x} \tag{6.5.4}$$

where $|k(x)| \geq k_0 = \text{constant} > 0$ for all x satisfying $0 \leq x \leq l$. Here we use the standard conventions that k is positive in heat transfer and negative in the elasticity applications, respectively.

By using (6.5.2), (6.5.3), and (6.5.4) to eliminate σ and Γ from (6.5.1), we find that there are two types of partial differential equations that arise in the physical problems described above. First, when (6.5.2) holds, u must satisfy

$$c(x, t) \frac{\partial u(x, t)}{\partial t} - \frac{\partial}{\partial x} \left[k(x) \frac{\partial u(x, t)}{\partial x} \right] = f(x, t),$$
$$0 < x < l, \quad t > 0 \tag{6.5.5}$$

whereas if (6.5.3) holds, we have

$$\rho(x) \frac{\partial^2 u(x, t)}{\partial t^2} - \frac{\partial}{\partial x} \left[k(x) \frac{\partial u(x, t)}{\partial x} \right] = f(x, t),$$
$$0 < x < l, \quad t > 0 \tag{6.5.6}$$

Equation (6.5.5) is a *parabolic* partial differential equation. Equations such as this describe diffusion processes, such as heat conduction. In the absence of source terms f and for homogeneous boundary conditions, the solutions to such equations have the property that they "decay" exponentially in time relative to some fixed initial data. In other words, if $f \equiv 0$ and $u \equiv 0$ on the boundary $x = 0$, $x = l$, and if $u(x, 0)$ is prescribed as some smooth function $\hat{u}(x) > 0$, then $u(x, t)$ will satisfy $u(x, t) < \hat{u}(x)$, for all $t > 0$.

Equation (6.5.6) is a *hyperbolic* partial differential equation. Equations such as this describe the propagation of waves. For example, if $f \equiv 0$, $u(x, 0) = \hat{u}(x)$, then these "initial data" will propagate through the domain at a finite speed determined by the functions ρ and k.

We shall describe a finite element analysis of the parabolic equation (6.5.5) and leave the study of (6.5.6) as an exercise. To simplify matters, we assume that the coefficient c is a smooth function of x, is independent of t, and that $c(x) \geq c_0 > 0$ for all x, c_0 being a positive constant. We also need to furnish boundary conditions which, for simplicity, we take to be

$$u(0, t) = u(l, t) = 0, \qquad t \geq 0$$

and an *initial condition* to start the evolution of the physical process at $t = 0$. This we assume to be of the form

$$u(x, 0) = \hat{u}(x), \qquad 0 < x < l$$

where $\hat{u}$ is a given smooth function of x. Finally, we are interested in studying this evolutionary process for only a finite period of time. Thus, we choose a final time T and study $u(x, t)$ for all t such that $0 \leq t \leq T$.

Our initial-value problem then takes the following form: find $u = u(x, t)$ such that

$$\left.\begin{array}{ll} c(x) \dfrac{\partial u(x, t)}{\partial t} - \dfrac{\partial}{\partial x}\left[k(x) \dfrac{\partial u(x, t)}{\partial x} \right] = f(x, t), & \begin{array}{l} 0 < x < l, \\ 0 < t \leq T \end{array} \\ u(0, t) = u(l, t) = 0, & 0 \leq t \leq T \\ u(x, 0) = \hat{u}(x), & 0 < x < l \end{array}\right\} \quad (6.5.7)$$

There are many ways to construct variational statements of problem (6.5.7). The most popular way, and the way we shall pursue here, is to consider t as a real parameter and to develop a *family of one-parameter variational problems* in t (for an alternative approach, see Exercise 6.5.2). By this, we mean the following: select an arbitrary smooth test function $v = v(x)$, *independent of t*, multiply the residual

$$r(x, t) = c(x)\, \partial u(x, t)/\partial t - \partial (k(x) \partial u(x, t)/\partial x)/\partial x - f(x, t)$$

by $v(x)$, integrate rv *only with respect to x*, and set this spatial weighted-average of rv equal to zero.

This process leads to the variational problem: find $u = u(x, t)$ such that, for every t, $0 \le t \le T$, $u(x, t) \in H_0^1(0, l)$ satisfies

$$\left.\begin{aligned}\int_0^l \left[c(x) \frac{\partial u(x, t)}{\partial t} v(x) + k(x) \frac{\partial u(x, t)}{\partial x} \frac{\partial v(x)}{\partial x} \right] dx \\ = \int_0^l f(x, t) v(x) \, dx \qquad \text{for all } v \in H_0^1(0, l)\end{aligned}\right\} \quad (6.5.8)$$

and which has the property that $u(x, 0) = \hat{u}(x)$, $0 < x < l$. Here $H_0^1(0, l)$ is the usual space of admissible functions with square-integrable derivatives for $0 < x < l$ and which vanish at $x = 0$ and at $x = l$.

The finite element approximation of (6.5.8) proceeds as in Chapter 2 with one major exception: the nodal values of the approximate solution u_h are now unknown functions of time. In other words, after constructing the basis functions $\phi_i = \phi_i(x)$, $i = 1, 2, \dots, N$, on a suitable mesh where N is the number of interior nodes, we take u_h to be of the form

$$u_h(x, t) = \sum_{j=1}^N u_j(t) \phi_j(x) \qquad (6.5.9)$$

where $u_j(t)$ is the value of u_h at node x_j and at time t,

$$u_h(x_j, t) = \sum_{i=1}^N u_i(t) \phi_i(x_j) = u_j(t)$$

The functions ϕ_i, of course, define a subspace H^h of $H_0^1(0, l)$.

The finite element approximation of (6.5.8) leads to a system of N ordinary differential equations in the N unknown functions $u_j(t)$ of the form

$$\mathbf{C} \frac{d\mathbf{u}(t)}{dt} + \mathbf{K}\mathbf{u}(t) = \mathbf{f}(t), \qquad 0 < t \le T \qquad (6.5.10)$$

$$\mathbf{u}(0) = \hat{\mathbf{u}}$$

wherein

$$\left.\begin{aligned}C_{ij} &= \int_0^l c(x) \phi_i(x) \phi_j(x) \, dx \\ K_{ij} &= \int_0^l k(x) \frac{d\phi_i(x)}{dx} \frac{d\phi_j(x)}{dx} \, dx \\ f_i(t) &= \int_0^l f(x, t) \phi_i(x) \, dx, \qquad 1 \le i, j \le N\end{aligned}\right\} \quad (6.5.11)$$

Now $\mathbf{u}$ is the $N \times 1$ vector of nodal values $u_j(t)$ and $\hat{\mathbf{u}}$ are interpolated values of $\hat{u}$ (i.e., an $N \times 1$ vector of nodal values of $\hat{u}(x)$). The matrix $\mathbf{C}$ is the so-called *capacitance* (or *mass matrix*) for the problem; K is the usual stiffness matrix for the steady-state problem, and f is the load vector, which is now time-dependent. Matrices C and K are sparse, narrowly banded, symmetric, and invertible. These matrices are generated from local element contributions just as in the case of steady-state problems (see Exercise 6.5.3).

By using the finite element method, we have succeeded in reducing the given initial-value problem (6.5.8) (or (6.5.7)) to the system of ordinary differential equations (6.5.10). Since we have not yet discretized the behavior of u_h in time, (6.5.10) is referred to as a *semidiscrete* finite element approximation. To obtain a *fully discrete* approximation, we must now introduce an approximation of the behavior of $u(t)$ in time. There are many ways to accomplish this. We shall only outline one of the simplest methods.

First, we partition the time domain $0 \leq t \leq T$ into M equal intervals of length $\Delta t = T/M$. At $t = 0$, the solution is known: $u(0) = \hat{u}$. To advance the solution in time from $t = n \, \Delta t$ to $(n + 1) \, \Delta t$, we use the forward finite-difference approximation,

$$\frac{d\mathbf{u}(n \, \Delta t)}{dt} \approx \frac{\mathbf{u}^{n+1} - \mathbf{u}^n}{\Delta t}, \qquad \mathbf{u}^n = \mathbf{u}(n \, \Delta t) \tag{6.5.12}$$

Then (6.5.10) leads to the algorithm

$$\mathbf{C}(\mathbf{u}^{n+1} - \mathbf{u}^n) = [-\mathbf{K}\mathbf{u}^n + \mathbf{f}^n] \, \Delta t$$

or

$$\mathbf{u}^{n+1} = (\mathbf{I} - \Delta t \mathbf{C}^{-1} \mathbf{K})\mathbf{u}^n + \Delta t \mathbf{C}^{-1} \mathbf{f}^n \tag{6.5.13}$$

Since $\mathbf{u}^0$ is known, $\mathbf{u}^1$ can be calculated using (6.5.13); this result is substituted into (6.5.13) to give $\mathbf{u}^2$, and so forth. In this way, we integrate the solution in time from $t = 0$ to $t = M \, \Delta t$ for any desired number of time steps. It should be noted that for a given mesh size h, limitations on the time-step size Δt are needed in order that this scheme be numerically stable. Other techniques, as well as limitations on the algorithm (6.5.13), are discussed in Volume III of this series (see also Exercise 6.5.5).

EXERCISES

6.5.1 (a) Develop a variational statement for an evolution problem characterized by the hyperbolic equation (6.5.6). Assume that $u(0, t) = u(l, t) = 0, t \geq 0$. Note that since second-order derivatives with respect to time appear in this problem, two initial conditions are needed,

$$u(x, 0) = \hat{u}(x), \qquad \frac{\partial u(x, 0)}{\partial t} = \hat{g}(x), \qquad 0 < x < l$$

(b) Describe the formulation of a finite element approximation of this hyperbolic problem. What differences are observed between this approximate formulation and that for the parabolic problem described in the text?

6.5.2 Describe an alternative variational formulation of problem (6.5.7) in which test functions v are selected which also depend upon t; $v = v(x, t), 0 < x < l$, $0 < t < T$. Show how one might formulate a finite element approximation of this problem. Compare the resulting system of equations obtained using this variational statement with those given in the text. What differences in methods for solving equations are suggested?

6.5.3 Consider problem (6.5.7) for the case

$$c = 1, \quad k = 1, \quad f = 0, \quad \text{and} \quad \hat{u}(x) - x(1 - x)$$

(a) Construct a finite element approximation of this problem using four elements of equal length and linear shape functions ψ_i^e. In particular, calculate the element mass and stiffness matrices $\mathbf{c}^e$ and $\mathbf{k}^e$ for each element,

$$c_{ij}^e = \int_{x_{e-1}}^{x_e} \psi_i^e \psi_j^e \, dx, \qquad k_{ij}^e = \int_{x_{e-1}}^{x_e} \psi_i^{e\prime} \psi_j^{e\prime} \, dx,$$

$$i, j = 1, 2, \quad e = 1, 2, 3, 4$$

[*HINT:* Many of these calculations have already been made in Chapter 1.]

(b) Formulate the global system of ordinary differential equations (6.5.10) for this problem.

(c) Using the algorithm (6.5.13), compute a fully discrete approximation for this problem for five time steps of length $\Delta t = 0.02$. Plot $u_h(x, t)$ as a function of x at the end of each time interval $n \, \Delta t, n = 0, 1, 2, 3, 4, 5$.

6.5.4 Discuss modifications in CODE1 of Chapter 3 that would make it possible to implement the method of Exercise 6.5.3.

6.5.5 The algorithm (6.5.13), often referred to as the *forward difference method*, is the result of using (6.5.12) in the evaluation of (6.5.10) at time $n \, \Delta t$.

(a) Instead of (6.5.12), use the backward difference scheme

$$\frac{d\mathbf{u}[(n + 1) \, \Delta t]}{dt} \approx \frac{\mathbf{u}^{n+1} - \mathbf{u}^n}{\Delta t}$$

to develop an "implicit" algorithm for the solution of (6.5.10).

(b)   Use $\mathbf{u}^{n+(1/2)} = \frac{1}{2}(\mathbf{u}^n + \mathbf{u}^{n+1})$ and the central difference approximation

$$\frac{d\mathbf{u}[(n + \frac{1}{2})\,\Delta t]}{dt} \approx \frac{\mathbf{u}^{n+1} - \mathbf{u}^n}{\Delta t}$$

to develop the "midstep" algorithm.

[*NOTE:* In contrast to the forward difference algorithm of (6.5.13), the backward and midstep difference algorithms are numerically stable for any value of Δt.]

INDEX

A

Accuracy, of finite element approximation, 36-39, 189, 197
 (*see* Errors)
Admissible functions (*see* Test function):
 for one-dimensional problems, 7
 for two-dimensional problems, 57, 147
 for three-dimensional problems, 223
 fourth-order problems, 231-32
A-posteriori estimates, 36
A-priori estimates, 36
Approximation:
 best Galerkin, 14-15
 of two-point problems by finite elements, 73-76
 of two-dimensional problems, 162, 165
 of time-dependent problems, 249
 of three-dimensional problems, 224
 of systems, 241
 interpretation of, 31-36
Area coordinates, 201-05
 definition, 201
 and nonlinear maps, 202

B

Backward difference scheme, 251
Banded matrix, 26
 elimination of, 118-21
 storage of, 118-21, 209
Bandwidth, definition of, 119
Basis functions:
 C¹-type, 232, 234, 236-37
 for Galerkin's method, 11-15, 59
 for model two-point problem, 15-23
 global, 33, 67, 69, 157, 234
 on rectangles, 159, 236-37
 on triangles, 154, 157, 237
 piecewise-constant, 22

piecewise-linear, 17, 20, 25, 30, 38, 67-68, 85, 154, 157
 piecewise-polynomial, 16
 piecewise-quadratic, 23
 Hermite, 233-37
 Lagrange polynomial, 67, 195, 204
Beam bending, 238-39
Bicubic rectangle, 237
Biharmonic equation, 235-36, 239
Body forces, 44
Boundaries:
 parametric definition of, 141-42, 193
 smoothness of, 132
Boundary conditions:
 calculation and flowchart of, 216
 clamped plate, 235
 compatability of, 229-30
 convective, 216
 essential, 43, 49, 55-56, 60, 81, 138, 140, 223, 231-32
 for biharmonic equation, 235
 for fourth-order problems, 229-30
 for general two-point problems, 42, 79-84
 in CODE1, 107-09
 in CODE2, 218-19
 natural (definition), 49, 138, 146, 231-32
Boundary condition data, input of, 104, 212-13
Boundary integrals, computation of, 191-92
Boundary, restriction of shape functions, 191, 197
Brick element, 226

C

C1 basis functions, 233-37
Capacitance matrix (*see* Mass matrix)
Chain rule:
 in area coordinates, 205
 in shape function transformation, 189
Charge, 44